AA004134

AIHealth 2025

The Second International Conference on AI-Health

March 9th –13th, 2025

Lisbon, Portugal

AIHealth 2025 Editors

Les Sztandera, Thomas Jefferson University, USA

Printed from e-media with permission by:

Curran Associates, Inc.
57 Morehouse Lane
Red Hook, NY 12571

Some format issues inherent in the e-media version may also appear in this print version.

Printed with permission by Curran Associates, Inc. (2025)

International Academy, Research, and Industry Association (IARIA)
412 Derby Way
Wilmington, DE 19810

Phone: (408) 893-6407
Fax: (408) 527-6351

petre@iaria.org

Additional copies of this publication are available from:

Curran Associates, Inc.
57 Morehouse Lane
Red Hook, NY 12571 USA
Phone: 845-758-0400
Fax: 845-758-2633
Email: curran@proceedings.com
Web: www.proceedings.com

Table of Contents

AIHealth 2025

Foreword

The Second International Conference on AI-Health (AIHealth 2025), held between March 9 - 13, 2025, covered topics blending Artificial Intelligence and health sciences.

Quality healthcare should be extended to all communities. Independent of how big and complex the healthcare systems are, physicians are under increasing time and workload pressures and spending less time with patients. The challenge to deliver high-quality healthcare against administrative burdens is big and increasing.

Healthcare facilities also produce great amounts of data and record high volumes of patient records information. This information is valuable and necessary to quality patient care. This information requires an enormousness effort (time, personnel) to be timely processed for prediction, evaluation and monitoring patients' health.

Artificial Intelligence (AI) comes to rescue in terms of accuracy, precision, rapidity and processing a large volume of data. AI-based health systems benefit for recent advances in sophisticated AI mechanisms for predicting patient health conditions (personalized, at large scale), producing useful analytics on varii patient health aspects, as well as monitoring and controlling patient under scrutiny.

We take here the opportunity to warmly thank all the members of the AIHealth 2025 Technical Program Committee, as well as the numerous reviewers. The creation of such a high quality conference program would not have been possible without their involvement. We also kindly thank all the authors who dedicated much of their time and efforts to contribute to AIHealth 2025.

Also, this event could not have been a reality without the support of many individuals, organizations, and sponsors. We are grateful to the members of the AIHealth 2025 organizing committee for their help in handling the logistics and for their work to make this professional meeting a success.

We hope that AIHealth 2025 was a successful international forum for the exchange of ideas and results between academia and industry and for the promotion of progress in the fields of AI and health sciences.

We are convinced that the participants found the event useful and communications very open. We also hope that Nice provided a pleasant environment during the conference and everyone saved some time for exploring this beautiful city

AIHealth 2025 Chairs:

AIHealth 2025 Steering Committee
Les Sztandera, Thomas Jefferson University, USA
Hesham H. Ali, University of Nebraska at Omaha, USA
Maura Mengoni, Università Politecnica delle Marche, Italy
Vitaly Herasevich, Mayo Clinic, USA
Amina Souag, Canterbury Christ Church University, UK
Hamid Shafie, VISKOOT, USA

AIHealth 2025 Publicity Chairs
Francisco Javier Díaz Blasco, Universitat Politècnica de València, Spain
Ali Ahmad, Universitat Politècnica de València, Spain

Sandra Viciano Tudela, Universitat Politecnica de Valencia, Spain
José Miguel Jiménez, Universitat Politecnica de Valencia, Spain

AIHealth 2025

Committee

AIHealth 2025 Steering Committee
Les Sztandera, Thomas Jefferson University, USA
Hesham H. Ali, University of Nebraska at Omaha, USA
Maura Mengoni, Università Politecnica delle Marche, Italy
Vitaly Herasevich, Mayo Clinic, USA
Amina Souag, Canterbury Christ Church University, UK
Hamid Shafie, VISKOOT, USA

AIHealth 2025 Publicity Chairs
Francisco Javier Díaz Blasco, Universitat Politècnica de València, Spain
Ali Ahmad, Universitat Politècnica de València, Spain
Sandra Viciano Tudela, Universitat Politecnica de Valencia, Spain
José Miguel Jiménez, Universitat Politecnica de Valencia, Spain

AIHealth 2025 Technical Program Committee

Alaa Abd-Alrazaq, Weill Cornell Medicine-Qatar, Doha, Qatar
Hesham H. Ali, University of Nebraska at Omaha, USA
Alireza Atashi, Tehran University of Medical Sciences, Iran
Michael Beigl, Karlsruhe Institute of Technology (KIT), Germany
Sid-Ahmed Berrani, Ecole Nationale Polytechnique, Algiers, Algeria
Elizabeth Borycki, University of Victoria, Canada
An Braeken, Vrije Universiteit Brussel, Belgium
Philippe Cinquin, CHUGrenoble Alpes, France
Marcos Cordeiro d'Ornellas, Universidade Federal de Santa Maria (UFSM) | Hospital Universitário (HUSM), Brazil
Manuel Domínguez-Morales, University of Sevilla, Spain
Sai Anvesh Durvasula, Parabole.ai, USA
Duarte Duque, 2Ai - School of Technology | IPCA, Portugal
Vitaly Herasevich, Mayo Clinic, USA
Haralampos Karanikas, University of Thessaly, Greece
Sarfraz Khokhar, Rasimo Systems, USA
Baihua Li, Loughborough University, UK
Sushil K. Meher, All India Institute of Medical Sciences, New Delhi, India
James Meng, Norwich Medical School | University of East Anglia, UK
Maura Mengoni, Polytechnic University of Marche, Italy
Daniela Micucci, University of Milano - Bicocca, Italy
George Mihalas, Victor Babes Univ. Med.&Pharm, Timisoara | Academy of Medical Sciences, Com. Medical Informatics & Data Protection, Romania
Kartik Palani, iManage / University of Illinois Urbana-Champaign, USA
Bhanu Prakash KN, Bioinformatics Institute | A*STAR, Singapore
Nadav Rappoport, Ben-Gurion University of the Negev, Israel

Stefano Rinaldi, University of Brescia, Italy
Floriano Scioscia, Polytechnic University of Bari, Italy
Hamid Shafie, VISKOOT, USA
Amina Souag, Canterbury Christ Church University, UK
Gro-Hilde Severinsen, Norwegian Center for e-health research, Norway
Jaideep Srivastava, University of Minnesota, USA
Dalibor Stanimirovic, University of Ljubljana, Slovenia
Les Sztandera Thomas Jefferson University, USA
Hamid Usefi, Memorial University, Canada
Madhurima Vardhan, University of Massachusetts, Lowell, USA
Pi-Yang Weng, National ChengChi University, Taiwan

Copyright Information

For your reference, this is the text governing the copyright release for material published by IARIA.

The copyright release is a transfer of publication rights, which allows IARIA and its partners to drive the dissemination of the published material. This allows IARIA to give articles increased visibility via distribution, inclusion in libraries, and arrangements for submission to indexes.

I, the undersigned, declare that the article is original, and that I represent the authors of this article in the copyright release matters. If this work has been done as work-for-hire, I have obtained all necessary clearances to execute a copyright release. I hereby irrevocably transfer exclusive copyright for this material to IARIA. I give IARIA permission or reproduce the work in any media format such as, but not limited to, print, digital, or electronic. I give IARIA permission to distribute the materials without restriction to any institutions or individuals. I give IARIA permission to submit the work for inclusion in article repositories as IARIA sees fit.

I, the undersigned, declare that to the best of my knowledge, the article is does not contain libelous or otherwise unlawful contents or invading the right of privacy or infringing on a proprietary right.

Following the copyright release, any circulated version of the article must bear the copyright notice and any header and footer information that IARIA applies to the published article.

IARIA grants royalty-free permission to the authors to disseminate the work, under the above provisions, for any academic, commercial, or industrial use. IARIA grants royalty-free permission to any individuals or institutions to make the article available electronically, online, or in print.

IARIA acknowledges that rights to any algorithm, process, procedure, apparatus, or articles of manufacture remain with the authors and their employers.

I, the undersigned, understand that IARIA will not be liable, in contract, tort (including, without limitation, negligence), pre-contract or other representations (other than fraudulent misrepresentations) or otherwise in connection with the publication of my work.

Exception to the above is made for work-for-hire performed while employed by the government. In that case, copyright to the material remains with the said government. The rightful owners (authors and government entity) grant unlimited and unrestricted permission to IARIA, IARIA's contractors, and IARIA's partners to further distribute the work.

Revolutionizing Prostate Cancer Diagnosis: An Integrated Approach for Gleason Grade Classification and Explainability

Anil B. Gavade
Dept., of Electronics and Communication Engineering
KLS Gogte Institute of Technology, Belagavi-590008, India
Email: abgavade@git.edu

Rajendra B. Nerli
Dept., of Urology, JN Medical College, KLE Academy of Higher Education and Research. Belagavi 590010, India.
Email: rajendranerli@yahoo.in

Shridhar C. Ghagane
KAHER'S Dr. Prabhakar Kore Basic Science Research Center, JNMC Campus, Belagavi-590010, India
Email: shridhar.kleskf@gmail.com

Les Sztandera*
Dept., Computer Information Systems, Thomas Jefferson University, Philadelphia, PA 19107, USA
Email: Les.Sztandera@jefferson.edu

Abstract—**Accurate grading of Prostate Cancer (PCa) is vital for effective treatment planning and prognosis. This study introduces an advanced framework for Gleason Grade (GG) classification, addressing challenges in accuracy, computational efficiency, and interpretability. Utilizing the SICAPv2 dataset, which contains annotated prostate biopsy Whole Slide Images (WSIs) graded from GG0 to GG5, the framework integrates cutting-edge machine learning and deep learning techniques. Feature extraction is performed using a custom-designed Variational Autoencoder (VAE) with a VGG16 backbone, chosen for its computational efficiency, while dimensionality reduction with Principal Component Analysis (PCA) optimally selects 50 features for classification. The classification pipeline combines machine learning models, including Support Vector Machines (SVM), logistic regression, and random forests, with custom Deep Neural Networks (DNNs). SVM with an Radial Basis Function (RBF) kernel achieved an accuracy of 84% following hyperparameter tuning, while a custom five-layer dense neural network incorporating dropout and batch normalization demonstrated superior performance with an accuracy of 94.6%. Explainable AI (XAI) techniques, such as SHapley Additive exPlanations (SHAP), gradient-weighted class activation mapping (Grad-CAM), and Local Interpretable Model-Agnostic Explanations (LIME), enhance model interpretability by providing insights into feature importance and aligning predictions with clinical expertise. This framework delivers a robust, scalable, and interpretable solution for automated GG classification, bridging the gap between advanced AI techniques and clinical application.**

Keywords- Cancer diagnosis; Dimensionality reduction; Explainable AI; Feature extraction; Gleason grade classification.

I. Introduction

Prostate cancer remains a significant global health issue, ranking among the leading causes of cancer-related mortality in men. The prostate gland [6], located below the bladder and comparable in size to a walnut, plays a crucial role in male reproductive health by producing seminal fluid. Clinical manifestations often include Lower Urinary Tract Symptoms (LUTS), haematuria, erectile dysfunction, and urinary retention.

Traditional diagnostic methods, such as Digital Rectal Examination (DRE), prostate-specific antigen (PSA) screening, and 12-core Transrectal Ultrasound (TRUS)-guided biopsy, exhibit notable limitations. Over-diagnosis rates can reach up to 45%, while clinically significant cancers may be missed in 30% of cases. These challenges underscore the necessity for advanced diagnostic techniques to effectively distinguish aggressive from non-aggressive cancer types [1-4].

The integration of WSI with Artificial Intelligence (AI) presents transformative potential in prostate cancer diagnostics, particularly for GG. High-resolution digital images of prostate biopsy samples are acquired through WSI scanners and undergo preprocessing steps, such as normalization and artifact removal, to enhance image quality [8]. AI-driven models segment tissue regions and extract significant histopathological features using deep learning techniques, including Convolutional Neural Networks (CNNs) and VAE [5].

Following feature extraction, AI models classify tissue patterns into respective GG, facilitating precise cancer grading. Post-classification validation ensures model robustness, while explainability tools such as SHAP, LIME, Grad-CAM, and Saliency Maps enhance transparency and interpretability. Figure 1 illustrates the developed AI pipeline, addressing critical diagnostic challenges by improving accuracy, efficiency, and consistency in GG assessment.

This framework integrates VAEs, XAI techniques, and preprocessing methods to enhance GG classification precision, support personalized clinical decisions, and improve PCa outcomes. In this paper, Section II covers the related work, Section III delves into the methods and materials, Section IV presents the results and discussions, and Section V provides the conclusion.

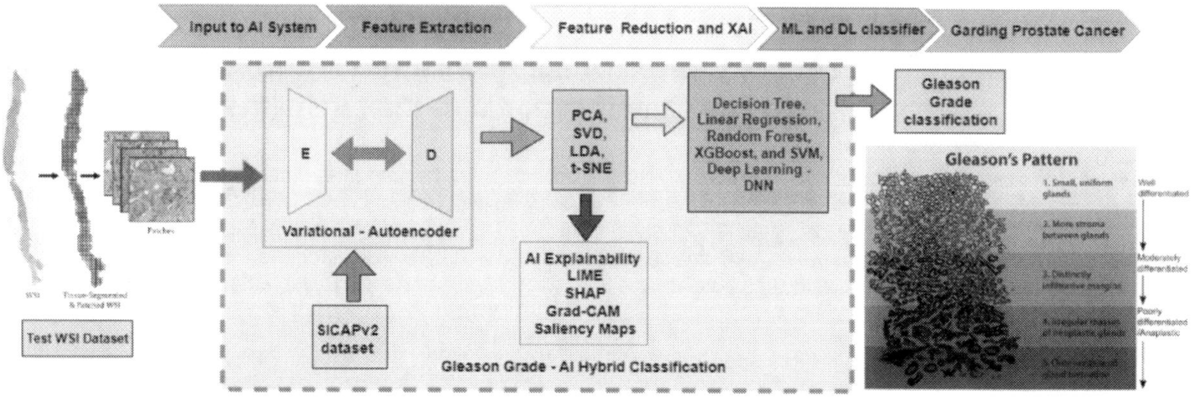

Figure 1. The block diagram shows a hybrid system for GG Group using WSIs, ensuring transparent and accurate PCa diagnosis and treatment with XAI techniques

II. Related Work

The integration of AI into the GG of PCa using WSI has brought about significant advancements in accuracy, consistency, and efficiency [9], [22]. Firjani et al. [10] laid the groundwork by achieving 100% accuracy in classifying prostate tissues into benign and malignant using a k-Nearest Neighbors (KNN) classifier on Diffusion-Weighted Imaging (DWI). Singhal et al. [11] improved segmentation and grading of PCa in WSIs of core needle biopsies with a DL model combining U-Net and Atrous Spatial Pyramid Pooling (ASPP) modules, achieving an accuracy of 89.4% and a quadratic-weighted kappa of 0.92. Azizi et al. [12] leveraged recurrent neural networks (RNN) on temporal enhanced ultrasound (TeUS) data, with Long Short-Term Memory (LSTM) networks achieving an accuracy of 0.93, an AUC of 0.96, a sensitivity of 0.76, and a specificity of 0.98. Bulten et al. [13] developed an automatic DL model for GG, attaining a quadratic Cohen's kappa score of 0.918 using biopsies. Tsuneki et al. [15] employed transfer learning to classify WSIs into adenocarcinoma and benign lesions, achieving a high ROC-AUC of up to 0.9873. Pati et al. [16] introduced WholeSIGHT, a weakly-supervised method for joint segmentation and classification, demonstrating a Dice coefficient of 0.76 on three public PCa WSI datasets. Müller et al. [17] presented DeepGleason, an open-source DNN system for automated GG, achieving a macro-averaged F1-score of 0.806, an AUC of 0.991, and an accuracy of 0.974. Hammouda et al. [18] proposed a multi-stage classification-based DL system for GG, achieving a precision of 0.92, recall of 0.89, and accuracy of 0.93 on 3,080 WSIs. Duenweg et al. [19] highlighted the impact of different WSI scanners on image quality, which significantly affects computational analysis performance, underscoring the need for standardized WSI scanner protocols. Mittmann et al. [20] developed an AI system for interpretable GG that mimics pathologist explanations, achieving a Dice score of 0.713 ± 0.003 using a dataset of 1,015 tissue microarray core images annotated by 54 pathologists. Belinga [11] proposed an AI-assisted system that improved GG accuracy and consistency, with a quadratically weighted Cohen's kappa of 0.872 compared to 0.799 without AI assistance, evaluated on 160 biopsies graded by 14 observers. Collectively, these studies underscore the transformative potential of AI and digital pathology in enhancing the diagnostic accuracy and consistency of GG in PCa.

III. Methods and Materials

Hybrid PCa GG uses a custom VAE with a pre-trained VGG-16 encoder for feature extraction and a two-layer Dense decoder for reconstruction. Trained on SICAPv2 datasets [5], [7], it ensures accurate GG classification and clinical relevance, as shown in Figure 2. To further optimize performance, we apply advanced feature reduction techniques, including PCA, Singular Value Decomposition (SVD), linear discriminant analysis (LDA), and t-distributed Stochastic Neighbor Embedding (t-SNE), ensuring dimensionality reduction while retaining critical data characteristics. The pipeline employs several state-of-the-art classifiers—Decision Tree, Random Forest, XGBoost, and SVM—which are fine-tuned via hyperparameter optimization to improve predictive accuracy. These classifiers are evaluated using performance metrics like accuracy, precision, recall, and F1-Score to ensure robust and reliable results. Furthermore, to enhance model transparency and interpretability, we incorporate XAI techniques. LIME offers local insights into individual predictions, SHAP quantifies global feature contributions, Grad-CAM visualizes critical regions in the images, saliency Maps highlight influential pixels, and feature Maps provide insights into the learning process at various layers. This comprehensive approach not only enhances the precision of GG but also supports transparency, ensuring the AI model is trustworthy for clinical use, and paves the way for more personalized PCa diagnosis and treatment strategies. The SICAPv2 dataset [22] (GG0–GG5) provided a robust benchmark for validating AI-driven PCa models.

Algorithm 1 VAE-Based Gleason Grade Classification

Input: Whole Slide Images (WSIs) $\{X_1, X_2, \ldots, X_N\}$, Labels $\{Y_1, Y_2, \ldots, Y_N\}$, Pre-trained Encoder E, Decoder D, Feature Reduction Method R, Classifier C.

Output: Predicted Gleason Grades $\{\hat{Y}_1, \hat{Y}_2, \ldots, \hat{Y}_N\}$.

Initialize the pre-trained VGG-16 encoder E and decoder D.
Construct Variational Autoencoder: $VAE = (E, D)$.
Train VAE on $\{X, Y\}$ to obtain latent representations:

$$Z = E(X), \quad Z \in \mathbb{R}^d.$$

Apply feature reduction R (e.g., PCA):

$$Z_R = R(Z), \quad Z_R \in \mathbb{R}^{d_r}.$$

Train the classifier C on $\{Z_R, Y\}$.
for $i = 1$ to N **do**
    Extract latent feature $z_i = E(X_i)$.
    Reduce dimensionality: $z_{i,R} = R(z_i)$.
    Predict Gleason Grade: $\hat{Y}_i = C(z_{i,R})$.
end for
Evaluate classification performance using:

$$\text{Accuracy} = \frac{\text{TP} + \text{TN}}{\text{TP} + \text{TN} + \text{FP} + \text{FN}},$$
$$\text{Precision} = \frac{\text{TP}}{\text{TP} + \text{FP}},$$
$$\text{Recall} = \frac{\text{TP}}{\text{TP} + \text{FN}},$$
$$\text{F1-Score} = 2 \cdot \frac{\text{Precision} \cdot \text{Recall}}{\text{Precision} + \text{Recall}},$$
$$\text{AUC-ROC} = \text{Area under the ROC curve}.$$

Figure 2. Implemenation algorithm VAE-Based Hybrid Algorithm for Gleason Grade Classification

IV. RESULTS AND DISCUSSION

This section presents the VAE-based hybrid pipeline's performance in GG classification, emphasizing key results, feature reduction, and XAI techniques.

A. *Feature Extraction by VAE and Feature reduction*

As shown in Table I, VGG-16 is the optimal feature extractor for GG classification, balancing quality, efficiency, and interpretability by extracting 512 compact and effective features. It avoids the redundancy seen in ResNet-50 and DenseNet-121, which produce 2048 and 1024 features, respectively. Despite DenseNet-121 being faster, VGG-16's moderate extraction time ensures reliability, minimizing overfitting and making it ideal for medical imaging.

TABLE I. FEATURE EXTRACTION

VAE with CNN as Encoder	**VAE Performance as Feature Extractor**				
	No. of features extracted from Model	***Feature Dimensions Before Flattening***	***Time taken by Model for FE***	***Time taken for feature Decoding***	***Time taken for PCA Transformation***
VGG-16	512	(None, 7, 7, 512)	63.06 sec	0.55 sec	0.05 sec
VGG-19	512	(None, 7, 7, 512)	65.50 sec	0.32 sec	0.01 sec
ResNet-50	2048	(None, 7, 7, 2048)	30.55 sec	0.28 sec	0.04 sec
DenseNet-121	1024	(None, 7, 7, 1024)	33.84 sec	0.32 sec	0.07 sec

Feature reduction performance in Table II indicates that various CNN models used as VAE encoders achieve similar dimensionality reduction to 50 features using PCA, SVD, and t-SNE. VGG-16, VGG-19, and ResNet-50 demonstrate comparable performance in feature reduction, with ResNet-50 extracting the highest number of features at 2048. DenseNet-121, extracting 1024 features, achieves the highest reduction with SVD, reducing features to 137. VGG-16 and VGG-19, with 512 features, consistently and efficiently reduce dimensionality while maintaining feature quality.

TABLE II. FEATURE REDUCTION

CNN Model As VAE Encoder	**Feature Reduction after VAE**				
	No. of features extracted from Model	***PCA***	***SVD***	***LDA***	***t-SNE***
VGG-16	512	50	100	1	2
VGG-19	512	50	93	1	2
ResNet-50	2048	50	103	1	2
DenseNet-121	1024	50	137	1	2

B. Feature Explainability

In Table IV, SHAP was the fastest, completing its task in 0.92 seconds while using only 1.47 MB of memory. LIME, although computationally intensive, required the highest memory at 9.97 MB. Grad-CAM stood out for its superior visual explanations, achieving a good balance with a runtime of 1.87 seconds and memory usage of 5.16 MB. Saliency maps provided a well-rounded performance, combining reasonable speed at 1.15 seconds with moderate memory usage of 6.74 MB. Figure 3 illustrates explainability techniques: (a) Significant contributions of 50 features to classification using XAI SHAP, (b) Grad-CAM heatmap for GG2 showing lower activation in cooler colors, indicating a lower likelihood of malignancy, and (c) Grad-CAM heatmap for GG4 displaying higher activation in warmer colors, highlighting regions significant for predicting malignancy.

TABLE III. EXPLAINABILITY OF FEATURE

XAI Technique	*Time (seconds)*	*Peak Memory Usage (MB)*
SHAP	0.9193	1.4740
LIME	1.4421	9.9731
Grad-CAM	1.8705	5.1574
Saliency Map	1.1513	6.7392

C. Machine Learnning classification

In Table IV, the performance metrics for various machine learning classification models are as follows: Decision Tree achieved accuracy, precision, recall, and F1-score of 0.47. Linear Regression showed consistent scores of 0.70 across all metrics. Random Forest performed better with scores of 0.78 across all metrics. XGBoost had moderate performance with scores of 0.72. SVM demonstrated the highest performance with accuracy and recall at 0.81, and precision and F1-score at 0.80.

TABLE IV. PERFORMANCE METRICS FOR VARIOUS ML CLASSIFICATION

Metric	ML Model				
	Decision Tree	*Linear Regression*	*Random Forest*	*XGBoost*	*SVM*
Accuracy	0.47	0.70	0.78	0.72	0.81
Precision	0.48	0.71	0.78	0.72	0.80
Recall	0.47	0.70	0.78	0.72	0.81
F1-Score	0.47	0.70	0.78	0.72	0.81

D. Machine Learnning classification with hyperparmeter tuning

In Table V, hyperparameter tuning significantly improved model performance. SVM showed the highest gains, with accuracy rising from 0.81 to 0.84, precision from 0.80 to 0.85, recall from 0.81 to 0.84, and F1-score from 0.81 to 0.84—improvements of 3.7% in accuracy and recall, and 3.8% in precision and F1-score. Random Forest improved from 0.78 to 0.81 in accuracy and recall, while XGBoost's accuracy rose from 0.72 to 0.76 and recall from 0.72 to 0.75. Decision Tree saw the largest improvement, with an 8.5% boost across all metrics (0.47 to 0.51). Linear Regression maintained consistent performance at 0.72. Overall, hyperparameter tuning enhanced all models, with SVM outperforming others in all metrics.

TABLE V. PERFORMANCE METRICS FOR VARIOUS ML CLASSIFICATION WITH HYPERPARMETER TUNING

Metric	ML Model hyperparameter				
	Decision Tree	*Linear Regression*	*Random Forest*	*XGBoost*	*SVM*
Accuracy	0.51	0.72	0.81	0.76	0.84
Precision	0.53	0.72	0.80	0.74	0.85
Recall	0.51	0.72	0.81	0.75	0.84
F1-Score	0.51	0.72	0.81	0.75	0.84

E. Deep learning – Deep neural network

In Table VI, we evaluated the performance of DNN models with different architectures. The DNN model with 3 Dense Layers achieved an accuracy of 0.79, precision of 0.81, recall of 0.79, and F1-score of 0.80. The performance improved with the DNN model featuring 5 Dense Layers, which achieved an accuracy of 0.89, precision of 0.90, recall of 0.89, and F1-score of 0.89. The model with 5 Dense Layers combined with Dropout and Batch Normalization demonstrated the best results, with an accuracy of 0.94, precision of 0.96, recall of 0.94, and F1-score of 0.95. The addition of Dropout and Batch Normalization led to a significant improvement, increasing accuracy by 5.6%, precision by 6.7%, recall by 5.6%, and F1-score by 6.7% compared to the model without these techniques. This indicates that Dropout and Batch Normalization played a crucial role in boosting the model's overall performance.

TABLE VI. DEEP LEARNIG – DNN MODEL

Metric	DL – DNN with 3 Dense Layers	DL – DNN with 5 Dense Layers	DL – DNN with 5 Dense Layers + Dropout & Batch Normalization
Accuracy	0.79	0.89	**94.6**
Precision	0.81	0.90	**96**
Recall	0.79	0.89	**94**
F1-Score	0.80	0.89	**0.95**

F. Comparative Analysis of ML and DL Techniques for Prostate Cancer Diagnostics

The performance comparison shows ML models like SVM achieving ~0.84 accuracy, while DL models excelled, with a 5-layer DNN using dropout and batch normalization achieving ~0.94; Figure 4 highlights that DL, combined with advanced regularization methods, offers superior accuracy and robustness in PCa GG classification.

V. CONCLUSION

In this study, a comprehensive framework for GG classification using WSI of PCa was developed by integrating DNN and ML models. VGG-16 was identified as the optimal feature extractor, offering a balance of feature quality and computational efficiency by extracting 512 features in 63.06 seconds. It outperformed DenseNet-121 and ResNet-50 in reducing redundancy and ensuring efficient dimensionality reduction through PCA, SVD, and t-SNE. Hyperparameter tuning enhanced ML performance, with SVM achieving the highest accuracy of 84%, while DL models incorporating dropout and batch normalization demonstrated significant improvements. A five-layer DNN achieved 94.6% accuracy, highlighting the effectiveness of regularization in preventing overfitting. A novel aspect of this research lies in the integration of XAI techniques to improve model interpretability. SHAP provided rapid, memory-efficient insights, while Grad-CAM delivered detailed visualizations, ensuring transparency in decision-making. LIME and Saliency Maps further contributed to understanding model outputs, underscoring the need for transparent AI in clinical settings. Future work will expand this framework to larger datasets and explore advanced neural architectures and XAI methods, aiming to develop scalable, interpretable, and clinically reliable AI models for PCa diagnostics. The implementation, tested on an open-access dataset, could benefit from additional testing on more benchmark and clinical datasets to enhance its clinical utility.

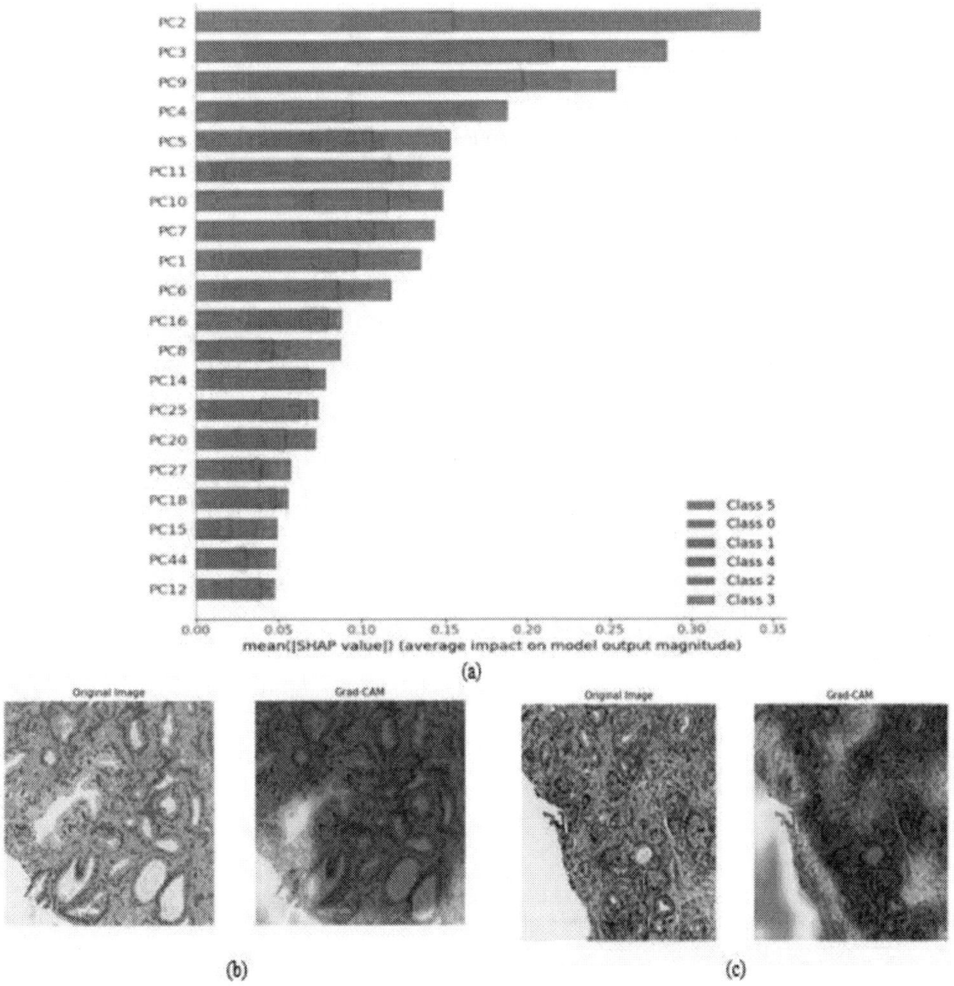

Figure 3. Comparison of Explainability Techniques for Prostate Cancer Gleason Grade Classification (a) Significant contributions of 50 features to classification using XAI SHAP (b) Grad-CAM heatmap for GG2 and (c) Grad-CAM heat map for GG4

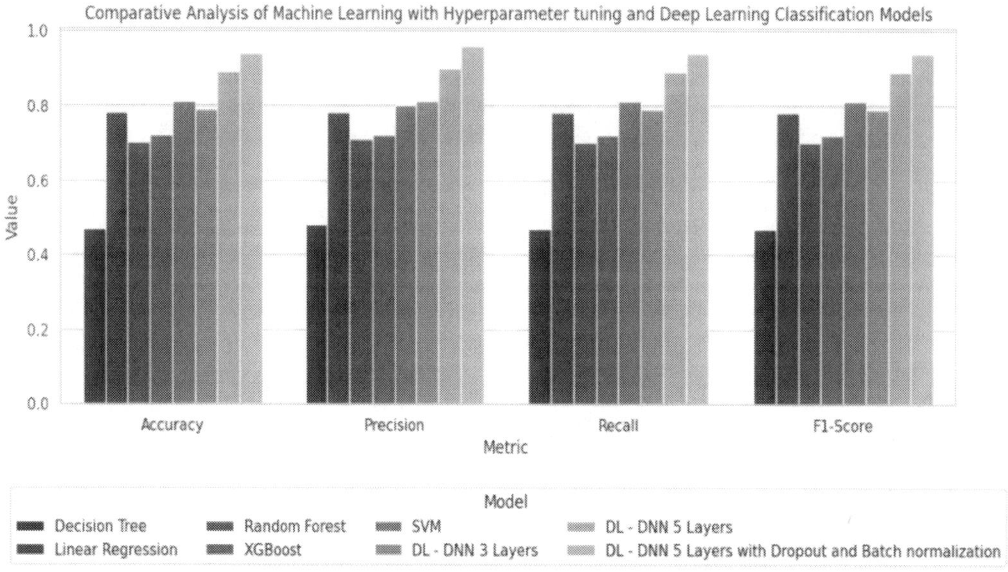

Figure 4. Evaluating ML and DL Models for Prostate Cancer Diagnostics: A Performance Insight

References

[1] American Cancer Society, "Cancer Facts & Statistics," accessed Jan. 15, 2025. https://www.cancer.org/research/cancer-facts-statistics.html.

[2] American Cancer Society, "Cancer Statistics Center," accessed Jan. 15, 2025. https://cancerstatisticscenter.cancer.org/#/.

[3] PathologyOutlines.com, "Prostate WHO Classification," accessed Jan. 15, 2025. https://www.pathologyoutlines.com/topic/prostateWHO.html.

[4] J. G. Kench et al., "WHO Classification of Tumours fifth edition: evolving issues in the classification, diagnosis, and prognostication of prostate cancer," *Histopathology*, vol. 81, no. 4, pp. 447-458, 2022.

[5] A. B. Gavade et al., "Innovative Prostate Cancer Classification: Merging Auto Encoders, PCA, SHAP, and Machine Learning Techniques," presented at *Int. Conf. Adv. Robot. Control Artif. Intell. (ARCAI 2024)*, Perth, Australia, Dec. 9–12, 2024. (unpublished).

[6] A. B. Gavade et al., "Automated diagnosis of prostate cancer using mpMRI images: A deep learning approach for clinical decision support," *Computers*, vol. 12, no. 8, p. 152, 2023.

[7] K. A. Gadad et al., "Beyond Single Models: Hybrid Approaches for Multiclass Cancer Identification," in *2024 3rd Int. Conf. Adv. Technol. (ICONAT)*, pp. 1-6, IEEE, 2024.

[8] R. B. Nerli et al., "Artificial Intelligence and Histopathological Diagnosis of Prostate Cancer," *J. Sci. Soc.*, vol. 51, no. 2, pp. 153-156, 2024.

[9] A. S. Balraj et al., "PRADclass: Hybrid Gleason Grade-Informed Computational Strategy Identifies Consensus Biomarker Features Predictive of Aggressive Prostate Adenocarcinoma," *Technol. Cancer Res. Treat.*, vol. 23, p. 15330338231222389, 2024.

[10] A. Firjani et al., "A diffusion-weighted imaging based diagnostic system for early detection of prostate cancer," *J. Biomed. Sci. Eng.*, vol. 6, no. 3, pp. 346, 2013.

[11] N. Singhal et al., "A deep learning system for prostate cancer diagnosis and grading in whole slide images of core needle biopsies," *Sci. Rep.*, vol. 12, no. 1, p. 1-11, 2022.

[12] S. Azizi et al., "Deep recurrent neural networks for prostate cancer detection: analysis of temporal enhanced ultrasound," *IEEE Trans. Med. Imag.*, vol. 37, no. 12, pp. 2695-2703, 2018.

[13] W. Bulten et al., "Automated deep-learning system for Gleason grading of prostate cancer using biopsies: a diagnostic study," *Lancet Oncol.*, vol. 21, no. 2, pp. 233-241, 2020.

[14] M. Tsuneki, M. Abe, and F. Kanavati, "A deep learning model for prostate adenocarcinoma classification in needle biopsy whole-slide images using transfer learning," *Diagnostics*, vol. 12, no. 3, p. 768, 2022.

[15] P. Pati et al., "Weakly supervised joint whole-slide segmentation and classification in prostate cancer," *Med. Image Anal.*, vol. 89, p. 102915, 2023.

[16] D. Müller et al., "DeepGleason: a System for Automated Gleason Grading of Prostate Cancer using Deep Neural Networks," *arXiv preprint arXiv:2403.16678*, 2024.

[17] K. Hammouda et al., "Multi-Stage Classification-Based Deep Learning for Gleason System Grading Using Histopathological Images," *Cancers*, vol. 14, no. 23, p. 5897, 2022.

[18] S. R. Duenweg et al., "Whole slide imaging (WSI) scanner differences influence optical and computed properties of digitized prostate cancer histology," *J. Pathol. Inform.*, vol. 14, p. 100321, 2023.

[19] G. Mittmann et al., "Pathologist-like explainable AI for interpretable Gleason grading in prostate cancer," *arXiv preprint arXiv:2410.15012*, 2024.

[20] A. Belinga, "AI-Enhanced Gleason Grading: A Comprehensive Approach," *arXiv preprint arXiv:2409.17122*, 2024.

[21] J. Silva-Rodríguez, "SICAPv2 - Prostate Whole Slide Images with Gleason Grades Annotations," *Mendeley Data*, V1, 2020. [Online]. Available: https://doi.org/10.17632/9xxm58dvs3.1 (accessed Dec. 30, 2024).

[22] J. P. Dominguez-Morales et al., "A systematic comparison of deep learning methods for Gleason grading and scoring," *Med. Image Anal.*, vol. 95, 103191, 2024.

Heavy Metals in Human Health and Pregnancy: How Data Analysis, Mining, and Modeling Present a Solution

Emma Frennborn, Khalil Rust, Les Sztandera
Thomas Jefferson University, Jefferson College of Life Sciences (JCLS)
Philadelphia, Pennsylvania, United States of America
Emails: Emma.Frennborn@students.jefferson.edu, Khalil.Rust@students.jefferson.edu, Les.Sztandera@jefferson.edu

Abstract— Essential heavy metals, such as zinc, copper, iron, and manganese, play important roles in many biological processes. Other heavy metals, like lead, arsenic, cadmium, and mercury, can displace essential heavy metals and disrupt vital biological processes. Heavy metal exposure can be particularly detrimental to pregnant women and their developing fetuses; however, little is known about the combinatory impact that simultaneous exposure to multiple heavy metals may have on fetal development. Procuring a better understanding of how these metals influence fetal development is a critical first step to addressing this concerning lack of knowledge. There are numerous databases and datasets in existence storing data on heavy metal levels in maternal and fetal blood, and novel studies are being done to expand this collection of data. Data mining techniques present a tool that could be used to close this gap of knowledge by revealing patterns in data not previously discovered. Research at Thomas Jefferson University aims to aid in closing this gap of knowledge. In the study, data such as heavy metal blood concentrations, pregnancy complications, health outcomes, and demographic information will be collected from mothers and newborns. Data mining strategies will then be used to develop models capable of discovering data patterns. If this modeling is successful, such an approach can be utilized by healthcare providers in the future to assess patient risk and provide early intervention for at-risk pregnant patients.

Keywords- data mining; data modeling; databases; heavy metals; neonatal health; maternal health.

I. INTRODUCTION

First, this paper will describe the critical roles that essential heavy metals play in human health, as well as how toxic metals can disrupt these processes. Also, the effects that toxic heavy metal exposure can have during pregnancy and on fetal development will be emphasized.

A. *Heavy Metals and Health*

Certain heavy metals are considered essential, as they act as co-factors in important enzymes necessary for maintaining biological homeostasis. Essential heavy metals include zinc, iron, copper, and manganese [1]. Zinc finger motifs, which require zinc, are common protein structures in Deoxyribonucleic Acid (DNA) binding proteins like transcription factors [2]. Iron is found in hemoglobin, which is involved in oxygen transport in red blood cells [3]. Copper and manganese are often found in enzymes involved in oxidation reduction (redox) reactions and both can be found in superoxide dismutases. Superoxide dismutases neutralize superoxide radicals formed in the mitochondria and act as antioxidants [3]. Considering the breadth of roles that heavy metals play in maintaining human health, disrupting the levels of these beneficial heavy metals in humans can have detrimental effects.

There are also other heavy metals known to cause adverse health effects. Lead, mercury, cadmium, aluminum, arsenic, and cobalt are some of the most well studied heavy metals associated with causing heavy metal toxicity [1]. Many of these toxic heavy metals displace essential heavy metals in important enzymes and thus prevent these enzymes from functioning properly. For example, cadmium and mercury can displace zinc in zinc finger motifs and can prevent transcription factors from binding to DNA [2]. Also, if the heavy metals in superoxide dismutases are displaced, there is an accumulation of reactive oxygen species leading to oxidative stress, which if severe enough, triggers apoptosis [3]. Toxic heavy metals also react with thiol groups, which are found on the amino acid cysteine, so any exposed cysteine on a protein has the potential to be altered by a toxic heavy metal [4]. Considering that zinc-finger motifs and thiol groups are found in many proteins, there is the potential for many biological functions to be disrupted when toxic heavy metals enter human cells, highlighting the importance of studying heavy metal exposure [4].

Although heavy metal exposure can occur at work, like in the case of miners and welders, the majority of the population is exposed to heavy metals from pollution [1] [4]. Heavy metals in water runoff from industrial pollution can enter crops, cattle, fish, and drinking water [4]. If contaminated food or water is consumed, heavy metals can enter the bloodstream, and in the case of pregnant women, can also be transferred to developing offspring [5]. The effects of heavy metals on developing fetuses can be especially damaging. There is substantial cell proliferation and differentiation during development which requires a great amount of energy and resources. Increased oxidative stress due to dysfunctional superoxide dismutases or decreased gene expression due to disrupted zinc finger motifs in DNA binding proteins can thus greatly damage the growth and development of fetuses [2][3][5].

Acute or chronic exposure to heavy metals during pregnancy can result in severe morbidities in mothers as well as in offspring. Exposure to heavy metals during pregnancy can lead to impaired growth and development in children. Heavy metals exposure during pregnancy is also associated with asthma, obesity, and hypertension [6]-[9]. Exposure to multiple heavy metals can also lead to deficiency of nutritionally essential metals with long-term consequences [6][10][11]. Maternal heavy metal exposure is also associated with decreased maternal thyroid hormones [12]. Elevated levels of lead in umbilical cord blood have been correlated with increased risk of preterm delivery, low birth weight, small head circumference, and low birth length [5]. Mercury is known to cross the placenta and can have negative effects on the developing nervous system of the fetus leading to cognitive delays, reduced attention span, memory issues, and motor skill impairments [13]. Cadmium can impair placental function, reducing the transfer of oxygen and nutrients to the fetus [14]. Arsenic also interferes with brain development and can also increase the risk of miscarriage and of developing chronic diseases such as diabetes or heart disease later in life [15]. These are just a few of the heavy metals that can have a negative impact on fetal development but more exist.

Simultaneous exposure to multiple heavy metals may produce a toxic effect that is either additive, antagonistic, or synergistic. However, the literature is scarce regarding the combined toxicity of heavy metals. Understanding how toxic metals may interact is necessary for our ability to predict the health outcome of a developing fetus and hopefully provide early intervention. There are many available datasets that exist reporting on the levels of heavy metals in maternal and fetal blood, but to our knowledge, data mining techniques that could be utilized to reveal patterns and create predictions based on the current data that exists have not yet been applied to this problem.

In this paper, we first aim to summarize data mining techniques that can be applied to biomedical research questions, including the gap of knowledge about the combinatory impact that simultaneous metal exposure has on fetal and maternal health. We also introduce research to be conducted at Thomas Jefferson University to address the heavy metals in neonatal health problem utilizing statistics and datamining techniques.

B. *Data Mining Techniques*

Many research-affiliated hospitals in the United States have established biobanks storing patient samples such as cord blood and maternal blood with associated databases. The Mayo Clinic Umbilical Cord Blood BioBank, Baylor College of Medicine's PeriBank, and the Magee Obstetric Maternal and Infant (MOMI) Database and Biobank are a few examples of large biobanks with databases [16][17]. As the data stored in these databases is sensitive, these databases are generally private and require permission to access. There are also several papers that have already been published with datasets which were used for analyzing heavy metal levels in maternal and fetal blood samples [18]-[23].

Considering that there are large datasets which already exist containing heavy metal levels from maternal and fetal blood, data mining could be utilized to reveal patterns not previously described about pre-existing data or could be used to predict future outcomes. There are a wide array of data mining methods available, which when used in conjunction, can be used to create an accurate modeling system. Many of these methods involve machine learning, which can be unsupervised or supervised [24].

Supervised machine learning involves randomly dividing the data into three subsets. One set is used to train the machine to establish the parameters. Then, a validation set is used to refine the model, and the test set is used to ensure the model performs as expected [24]. A Bayesian Network is a supervised machine learning model that could be used in the context of the heavy metal problem to profound effect. Bayesian Networks are probabilistic graphical models that represent a set of variables as a Directed Acyclic Graph (DAG) [25]. Nodes in the model represent random variables while connections between nodes represent conditional probabilities. This model can be used to calculate the probability of a certain outcome given several different interacting variables [25]. An approach like this could help to elucidate the potential synergistic, additive and antagonist properties of multiple heavy metals, and predict outcomes based on the level of exposure of individual metals. Supervised machine learning requires that data is labeled and is an approach used to train a model [24]. This means there is a known output expected when analyzing the data. If the goal is to reveal patterns in the data, in which the output is unknown and undefined, unsupervised machine learning is the appropriate approach [24].

Unsupervised machine learning models, also known as descriptive models, are often used to find patterns that describe data and can be interpreted by humans [24]. These models achieve this by clustering data into categories based on the similarity between objects in the dataset. Unsupervised data mining is exploratory in nature and can lead to the discovery of unknown patterns or relationships in data [26]. An example of this mining technique would be an association rule-based analysis, also known as market based analysis. Association rules can discover correlations between items in substantial amounts of data, and this technology is often used in health care settings to determine associations between joint effects of disease risk factors and combinations of other risk factors [26]. This is done in a 2-step process: 1) all high frequency items in the data set are listed and 2) frequent association rules are generated based on these high frequency items [26]. Such an approach would lend itself to the heavy metal problem quite nicely. Determining association rules between the various heavy metals and their effect on human health or fetal development could clarify how their known effects interact with each other and even aid in the discovery of new combinatorial effects that were previously uncharacterized.

Section 2 of this paper will describe a proposed study to be performed at Thomas Jefferson University in regard to this topic. Specifically, section 2A will describe the data collection process, while section 2B will go over the proposed analysis strategy. Section 3 and the acknowledgement section will conclude the paper.

II. PROPOSED RESEARCH

The following section will describe the data collection process.

A. *Data Collection*

In an Institutional Review Board (IRB) approved study at Thomas Jefferson University, 107 pregnant women will be enrolled. Consent will occur both in the outpatient setting prior to delivery and on admission to Labor and Delivery (L & D). L&D is a specialized unit that provides care to pregnant women during labor, childbirth, and the immediate postpartum period. Maternal blood samples will be collected on admission to L & D and the cord blood at delivery. Blood samples will be centrifuged, serum collected, and frozen at -80°C. Serum samples will be analyzed by mass spectrometry for 25 metals (Na, K, Mg, Ca, Zn, Se, Cu, Li, Co, NI, Tl, Al, Cr, Sr, Cd, Ba, Be, V, Fe, As, Mo, Pb, Ag, Mn, and U). The following clinical data will be collected: maternal age, race/ethnicity, insurance, zip code, Body Mass Index (BMI), medical morbidities (asthma, hypertension, diabetes), pregnancy complications (preeclampsia, preterm birth, fetal growth restriction), maternal anemia, prior full term or preterm delivery, delivery outcomes (delivery mode, gestational age, birthweight, neonatal sex) and neonatal outcomes (duration of hospital stay, admission to the Neonatal Intensive Care Unit (NICU). The World Health Organization guidelines will be adhered to for postnatal care, including routine postpartum evaluation of all women and infant pairs at 3 days, 1-2 weeks, 6 weeks, and 12 weeks postpartum. The next section will detail the methodology that will be used for data analysis.

B. *Data Analysis*

Data will be analyzed through International Machines Corporation (IBM) Analytics software based on the project outline that is divided into distinct, but interlocked research goals. First, we aim to develop a database of blood samples from umbilical cords and mothers that reflect the maternal toxic elements and their potential transfer, as well as mother and newborn nutritional status. Secondly, we aim to develop a data analytics model to discover and prioritize data patterns. IBM Analytics software will be utilized to calculate power analysis for all statistical analysis to be undertaken in the newborn and mothers' blood samples study. We estimate that, for testing whether the mother's blood at the delivery will predict the toxicity level transferred to a newborn as well as whether the relation is being affected through the metals' mixture composition for the toxic materials, in order to achieve the value of power 0.95 as well as the medium size effect in regression analysis, the sample size required to be selected is around 107. Thus, it can be inferred that Power Analysis for the method of regression will help in stating the exact size of the sample on the basis of the research questions to demonstrate statistical significance. IBM Analytics software allows for multiple analysis of mean group differences and variance. The Variance Components procedure, for mixed-effects models, estimates the contribution of each random effect to the variance of the dependent variable. By calculating variance components, we will determine where to focus attention to reduce the variance in the computational models. We intend to explore four different methods for estimating the variance components: minimum norm quadratic unbiased estimator (MINQUE), analysis of variance (ANOVA), Maximum Likelihood (ML), and Restricted Maximum Likelihood (REML). If the ML method or the REML method is used, an asymptotic covariance matrix table is also displayed. Other available output includes an ANOVA table and expected mean squares for the ANOVA method and an iteration history for the ML and REML methods. WLS Weight will allow us to specify a variable used to give observations different weights for a weighted analysis to compensate for variation differences. ANOVA and MINQUE do not require normality assumptions. ML and REML require the model parameter and the residual term to be normally distributed. In terms of Data Management and Quality Control Mechanisms, three standard Jefferson security procedures based on Jefferson Information Systems and technology (IS&T) / Information Security questionnaires will be utilized to review data hosting to assure compliance with applicable security controls. The details of these internal procedures cannot be disclosed publicly as to protect proprietary information belonging to Thomas Jefferson University. Thomas Jefferson University recognizes interoperability as crucial to the sharing of research data and resources to promote efficiency in research, and utilizes standards articulated by the Jefferson Research Integrity, Conduct and Compliance Office. Software applications are hosted on servers with networked storage located at a data center, providing data security and disaster recovery services. Jefferson employs best practices with regard to data privacy and security, complying with the Common Rule, HIPAA, as well as state regulations.

In terms of Data Analytics, we aim at developing models to discover and prioritize data patterns to provide information and actionable knowledge to both medical practitioners as well as public health policy decision makers. Data Analytics will be used: 1. To explore data to find new patterns and relationships (data mining); 2. To evaluate and test previous decisions (randomized controlled experiments, multivariate testing); 3. To explain why a certain outcome happened (statistical analysis, descriptive analysis); and 4. To venture into the future (forecast) results (predictive modeling, predictive analytics). All four research avenues capture very well the significance and impact of Data Analytics. It could be summarized as a leading theme for the whole research proposal: "In God we trust, all others bring data, especially in maternity health care, and its impact on future healthy cities."

The data from the proposed research will help in identifying pregnant women at risk for developing heavy metals toxicity and the deficiency of essential nutrients. Identifying deficiency of nutritionally essential metals in pregnant women and their newborn and supplementation may improve pregnancy outcomes, as well as improve growth and development in children and prevent long-term morbidities. The significance of the proposed research is accentuated by environmental impact on maternal/child health through measuring maternal heavy metal exposure and fetal transfer. Although the acute and chronic effects are known for some metals, little is known about the health impact of mixtures of toxic elements.

III. CONCLUSION

Heavy metals play a complex role in human health. Harmful metals are known to interact with essential metals in a competitive way, but their combinatorial effects have not been sufficiently studied. An understanding of these combinatorial effects is important as individuals are often exposed to many of these harmful metals at once, due to the impact of industrial pollution. Fortunately, there is a growing collection of data concerning the impact of these metals on pregnant mothers and newborns. Data mining technologies present an efficient and productive method to utilize these databases. Proper use of datamining techniques used in conjunction with the collected data on these metals could elucidate the additive, synergistic, and antagonistic effects of these metals, thereby filling a gap of knowledge. As such, exploration of this topic would be of immense importance. At Thomas Jefferson University, we aim to collect data on maternal and neonatal heavy metal blood levels and health outcomes to add to the data already in existence. Then, data mining techniques will be used to develop data modeling systems capable of revealing patterns in this data not previously reported. Understanding the connection between health outcomes and heavy metal blood levels in both mothers and developing fetuses could allow clinicians to better predict pregnancy and delivery complications and thus could provide early intervention, if needed, to prevent complications. These healthcare goals can only be achieved once the data available on heavy metal levels during pregnancy/delivery are analyzed and modeled.

ACKNOWLEDGEMENT

The authors would like to thank Dr. Les Sztandera of Thomas Jefferson University for his mentorship and guidance. Also, the authors gratefully acknowledge the travel funding received from the Jefferson Global Health Initiatives Council (GHIC) Global Office, which made presenting this work at the IARIA AI Health 2025 conference possible.

REFERENCES

[1] P. B. Tchounwou, C. G. Yedjou, A. K. Patlolla, and D. J. Sutton, "Heavy metal toxicity and the environment," *Exp. Suppl.*, vol. 101, pp. 133-164, 2012.

[2] A. Hartwig, "Zinc finger proteins as potential targets for toxic metal ions: differential effects on structure and function," *Antioxid. Redox Signal.*, vol. 3, no. 4, pp. 625-634, Aug. 2001.

[3] M. Jaishankar, T. Tseten, N. Anbalagan, B. B. Mathew, and K. N. Beeregowda, "Toxicity, mechanism and health effects of some heavy metals," *Interdiscip. Toxicol.*, vol. 7, no. 2, pp. 60-72, Jun. 2014.

[4] D. Witkowska, J. Słowik, and K. Chilicka, "Heavy metals and human health: Possible exposure pathways and the competition for protein binding sites," *Molecules*, vol. 26, no. 19, p. 6060, 2021.

[5] S. S. Zinia, K. H. Yang, E. J. Lee, M. N. Lim, J. Kim, W. J. Kim, and the Ko-CHENS Study Group, "Effects of heavy metal exposure during pregnancy on birth outcomes," *Sci. Rep.*, vol. 13, no. 1, p. 18990, Nov. 2023

[6] I. Karakis, D. Landau, M. Yitshak-Sade, et al., "Exposure to metals and congenital anomalies: a biomonitoring study of pregnant Bedouin-Arab women," *Sci. Total Environ.*, vol. 517, pp. 106-112, 2015.

[7] J. M. Madrigal, V. Persky, A. Pappalardo, and M. Argos, "Association of heavy metals with measures of pulmonary function in children and youth: Results from the National Health and Nutrition Examination Survey (NHANES)," *Environ. Int.*, vol. 121, pt. 1, pp. 871-878, 2018.

[8] I. Al-Saleh, N. Shinwari, A. Mashhour, and A. Rabah, "Birth outcome measures and maternal exposure to heavy metals (lead, cadmium, and mercury) in Saudi Arabian population," *Int. J. Hyg. Environ. Health*, vol. 217, no. 2-3, pp. 205-218, 2014.

[9] S. F. Farzan, C. G. Howe, Y. Chen, et al., "Prenatal lead exposure and elevated blood pressure in children," *Environ. Int.*, vol. 121, pt. 2, pp. 1289-1296, 2018.

[10] I. Karakis, D. Landau, R. Gat, et al., "Maternal metal concentration during gestation and pediatric morbidity in children: an exploratory analysis," *Environ. Health Prev. Med.*, vol. 26, no. 1, p. 40, 2021.

[11] G. O. Arinola, A. Dutta, O. Oluwole, C. O. Olopade, "Household air pollution, levels of micronutrients and heavy metals in cord and maternal blood, and pregnancy outcomes," *Int. J. Environ. Res. Public Health*, vol. 15, no. 12, 2018.

[12] X. Sun, W. Liu, B. Zhang, et al., "Maternal heavy metal exposure, thyroid hormones, and birth outcomes: A prospective cohort study," *J. Clin. Endocrinol. Metab.*, vol. 104, no. 11, pp. 5043-5052, 2019.

[13] C. S. Oliveira, P. A. Nogara, D. M. P. Ardisson-Araújo, M. Aschner, J. B. T. Rocha, and J. G. Dórea, "Neurodevelopmental effects of mercury," *Adv. Neurotoxicol.*, vol. 2, pp. 27-86, 2018. doi: 10.1016/bs.ant.2018.03.005. Epub May 24, 2018.

[14] J. L. Young and L. Cai, "Implications for prenatal cadmium exposure and adverse health outcomes in adulthood," *Toxicol. Appl. Pharmacol.*, vol. 403, p. 115161, Sep. 2020.

[15] N. Y. Ortiz-Garcia, et al., "Maternal exposure to arsenic and its impact on maternal and fetal health: A review," *Cureus*, vol. 15, no. 11, p. e49177, Nov. 2023.

[16] S. N. Caritis, "Magee obstetric maternal & infant (MOMI) database and Biobank: Magee-Womens Research Institute & Foundation," *Magee*..

[17] K. M. Antony, et al., "Generation and validation of a universal perinatal database and biospecimen repository: PeriBank," *J. Perinatol.*, vol. 36, no. 11, pp. 921-929, Nov. 2016.

[18] I. B. Goldberg, et al., "Early pregnancy metal levels in maternal blood and pregnancy outcomes," *Sci. Rep.*, vol. 14, no. 1, p. 27866, Nov. 2024

[19] K. Kot, et al., "Interactions between 14 elements in the human placenta, fetal membrane and umbilical cord," *Int. J. Environ. Res. Public Health*, vol. 16, no. 9, p. 1615, 2019.

[20] I. Al-Saleh, N. Shinwari, A. Mashhour, G. E. D. Mohamed, A. Rabah, "Heavy metals (lead, cadmium, and mercury) in maternal, cord blood and placenta of healthy women," *Int. J. Hyg. Environ. Health*, vol. 214, no. 2, pp. 79-101, 2011.

[21] T. Takatani, et al., "Association between maternal blood or cord blood metal concentrations and catch-up growth in children born small for gestational age: an analysis by the Japan environment and children's study," *Environ. Health*, vol. 23, no. 1, p. 18, Feb. 2024.

[22] J. Grzesik-Gąsior, J. Sawicki, A. Pieczykolan, and A. Bień, "Content of selected heavy metals in the umbilical cord blood and anthropometric data of mothers and newborns in Poland: preliminary data," *Sci. Rep.*, vol. 13, no. 1, p. 14077, Aug. 2023.

[23] H. Ma, et al., "Maternal and cord blood levels of metals and fetal liver function," *Environ. Pollut.*, vol. 363, pt. 2, p. 125305, 2024.

[24] W. T. Wu, Y. J. Li, A. Z. Feng, et al., "Data mining in clinical big data: the frequently used databases, steps, and methodological models," *Mil. Med. Res.*, vol. 8, p. 44, 2021.

[25] J. Puga, M. Krzywinski, and N. Altman, "Bayesian networks," *Nat. Methods*, vol. 12, pp. 799–800, 2015.

[26] I. Yoo, et al., "Data mining in healthcare and biomedicine: a survey of the literature," *J. Med. Syst.*, vol. 36, no. 4, pp. 2431-2448, Aug. 2012. Epub May 3, 2011.

Recent Advances in Machine Learning for Log File-Based PSQA for IMRT and VMAT

Kellin M. DeJesus
Cell Biology and Regenerative Medicine
Thomas Jefferson University
Philadelphia, Pennsylvania, USA
e-mail: kmd119@students.jefferson.edu
Leon Dunn
Medical Physics
Genesis Care
Victoria, Australia
e-mail: info@efilmqa.com
David Thomas
Medical Physics
Thomas Jefferson University
Philadelphia, Pennsylvania, USA
e-mail: david.thomas2@jefferson.edu
Les Sztandera
Computer Science
Thomas Jefferson University
Philadelphia, Pennsylvania, USA
e-mail: les.sztandera@jefferson.edu

Abstract—**The paper addresses the critical need for a faster and more efficient approach to Patient-Specific Quality Assurance (PSQA) in radiation therapy. The accuracy of PSQA is crucial for the safety of radiation therapy, particularly with complex procedures like Intensity-Modulated Radiation Therapy (IMRT) and Volumetric-Arc Radiation Therapy (VMAT). Traditional phantom-based methods, while effective, are time-consuming and fail to account for patient-specific variability and real-time treatment adjustment. To address these limitations, alternative strategies leveraging trajectory log files—automatically recorded during treatment—have emerged as promising tools for PSQA. In recent years, the application of machine learning and deep learning algorithms to trajectory log files has been increasingly studied in literature. These algorithms have shown notable progress in predicting PSQA outcomes and detecting errors, though further development is required before they can be fully integrated into clinical practice. By surveying key studies, the paper highlights the potential of algorithms such as support vector machines, tree-based methods, and convolutional neural networks to enhance the efficiency and accuracy of log file-based PSQA. The findings underscore the promise of these techniques in replacing traditional methods while addressing current challenges to pave the way for clinical integration.**

Keywords-deep learning; machine learning; quality assurance; volumetric-arc radiation therapy; intensity-modulated radiation therapy.

The American Cancer Society has estimated over 2 million new cases of cancer in 2024 [1]. About 50% of all cancer patients are expected to receive radiotherapy at some point during treatment [2]. The proportion of radiotherapy patients receiving Intensity-Modulated Radiotherapy (IMRT) and Volumetric Modulated Arc Therapy (VMAT) has steadily increased over time from 22% in 2004 to 57.8% in 2017 [3]. IMRT and VMAT are routine but complex cancer treatment modalities that require time-consuming Quality Assurance (QA) measures. Log file-based Patient-Specific Quality Assurance (PSQA) has been proposed as an alternative method that can be performed in real-time on a fraction-by-fraction basis [4][5][6]. Studies comparing log file-based PSQA have identified differences between log file recordings and actual behavior of machines during treatment, however, several mitigation strategies have been proposed [4][7][8]. These studies have given new insights into the potential for more efficient PSQA, however, they have been limited by small cohort size.

Machine learning, and by extension deep learning, have rapidly gained traction as essential tools for advancing healthcare [9][10][11]. Machine learning can process and analyze large, complex datasets to identify patterns and make predictions that can be implemented to improve patient outcomes, increase treatment efficiency, and aid in clinical decision-making. Machine learning algorithms can automate time-consuming tasks. This can reduce the workload on medical professionals, reduce waiting times, and mitigate the risks of human error. Unlike traditional strategies for automation that are static after their implementation, these algorithms can evolve over time with additional data. Updates are made constantly to maintain or improve accuracy [12]. This is specifically advantageous in fields, such as radiation therapy, where advancements are rapid, and techniques are constantly changing [13][14][15][16].

The following paper thus endeavors to give a brief but comprehensive overview of the current status of machine learning for log-file based PSQA measures. This paper is structured as follows: Section 2 provides the theoretical context for log-file based PSQA. Section 3 explores the various applications of machine learning and deep learning models for PSQA. Section 4 discusses future directions and concludes with final remarks.

I. BACKGROUND

We will provide an overview of the theories behind the use of log files for PSQA and the theory for the most successful machine learning algorithms to date.

A. Log File-Based PSQA

IMRT is of particular value when treating tumors with complex or concave shapes, especially those located near radiosensitive normal tissues. It uses a computer-controlled linear accelerator (linac) that can rotate around the patient on a gantry. This process excites electrons via microwave technology, which then collide with a heavy metal target to produce high-energy x-rays. These beams are shaped by the Multileaf Collimator (MLC) as they exit the machine. The intensity of each beam segment, MLC shaping, and gantry rotation are all determined using 3D imaging prior to treatment. The precision in dose delivery provided by IMRT allows for the irradiation of the tumor while sparing nearby healthy tissues, making it especially useful for tumors located near critical organs [17].

Due to the complexity of IMRT and VMAT treatment plans, each patient's treatment plan requires inversely optimized planning. Before treatment begins, these plans are often measured on the linac using detector arrays. The added complexity necessitates additional QA measures to ensure patient safety in clinical settings [18][19]. Confirmations of machine performance and patient treatment plan accuracy are essential. These verifications include assessing patient positioning, machine mechanical accuracy, dose distribution, and beam geometry. Given the complex and highly variable nature of each treatment plan, PSQA is required [20].

Currently, IMRT and VMAT treatment plans undergo physical measurements of the plan parameters before treatment begins to ensure the machine delivers the intended dose. However, these measurements are often done in advance and may not account for real-time deviations that may occur before the treatment begins. As such, the potential for mechanical changes resulting in dose discrepancies between PSQA and actual treatment remains a concern. The most common procedure involves recalculating the dose distribution of a patient's treatment plan onto a suitable phantom. The dose distribution is then measured using various devices, such as film, ion chambers, diode arrays, or Electronic Portal Imaging Devices (EPIDs). Differences between the measured and planned dose distributions are quantified using gamma analysis, as outlined in American Association of Physicists in Medicine Task Groups (AAPM TGs) 119 and 218. These guidelines recommend that over 90% of measured points should fall within a 3% dose difference and a 2mm distance-to-agreement (DTA) [21][22]. However, this process is time and labor-intensive, often requiring after-hours work to avoid interrupting treatment schedules. Additionally, there is ongoing debate over the efficacy of these methods, particularly regarding their robustness and ability to detect potential failure modes [6][23][24].

Log file-based PSQA offers an alternative to traditional methods by utilizing automatically generated log files from radiation treatment machines to verify the accuracy of treatment plans. These log files capture data such as radiation output, MLC positions, gantry and couch positions, beam angles, and timing information. This data can then be compared to the treatment plan to identify potential errors [25]. Log files, TPS files, and Mean Complexity Scores (MCS) have been used to develop prediction models for Gamma Passing Rate (GPR), a key metric in PSQA [26]. Recent studies have indicated discrepancies between the recorded data and the actual performance of the machine, particularly in terms of MLC positioning [26][27][28]. However, since log files are generated by the linac, they do not detect mechanical miscalibrations, such as incorrect leaf positioning. Moreover, they cannot account for low plan quality or errors originating from the treatment planning system (TPS). To mitigate these limitations, enhanced QA procedures specifically for the linac, combined with more sensitive machine QA tools, are recommended to ensure MLC accuracy [26].

B. Machine Learning and IMRT/VMAT

Treatment log files record various parameters of radiation delivery, such as MLC position, dose rates, beam angles, and gantry positions in real-time during the course with recordings taken every few milliseconds [29]. As highly structured, real-time, and extensive data capture, these files would be particularly difficult to analyze manually. Log files are thus particularly well-suited to machine learning algorithms for pattern recognition and error prediction. Models range from simple classification techniques to complex deep learning algorithms. The most successful models in the literature include Support Vector Machines (SVMs), tree-based algorithms , and Artificial Neural Networks (ANNs).

SVMs are effective for classification tasks for log file-based PSQA. They can distinguish between compliant and non-compliant treatment sessions by setting predefined acceptable ranges for discrepancies between planned and delivered values for parameters within the log file, such as dose rate, MLC positions, and beam angles. This allows for quick identification of errors as they occur so that a clinician can be alerted. However, SVM is limited to cases where there are clear distinctions between compliant and non-compliant values. SVM is also sensitive to noise and outliers and is not well suited for multi-class tasks [30].

Tree-based algorithms are non-parametric and based on hierarchical, tree-like structures. Each tree is made up of nodes that represent decisions based on feature values. The branches represent possible outcomes or decisions. They are well-suited for non-linear relationships between features and can partition the feature space in more complex ways than linear models. Tree-based machine learning models include

Random Forest (RF), Gradient Boosting, and Extreme Gradient Boosting (XGBoost) algorithms [31][32][33].

RF models can leverage many decision trees to map the involvement of multiple interacting features to identify more subtle discrepancies between expected and delivered values. It can detect complex relationships within the treatment data that would not be as apparent with simpler methods such as SVM. Due to the ensemble nature of the algorithm, RFs are difficult to interpret and feature importance scores are only rough approximations. They can show bias toward categorical features with many levels. RFs also require a lot of optimizations for hyperparameter tuning [31].

Gradient Boosting uses decision trees as its base and adjusts instance weights with each iteration by fitting new predictors to errors in the preceding iteration. Individual decision trees are differentiated by a different subset of features to select the best split. Each new tree accounts for the errors of the preceding ones. This approach can be slow to train and is prone to overfitting [32]. XGBoost builds upon the gradient boosting algorithm by including L1 (Lasso) and L2 (Ridge) regularization to prevent overfitting [34][35]. It also grows trees with a depth-first approach and can train trees in parallel, which increases the speed of training. Although these two models are less prone to overfitting than RF, they do still pose some risk of overfitting. They also exhibit hyperparameter sensitivity and require careful tuning, especially for large datasets. Like other tree-based models, they both struggle with extrapolation beyond the training dataset [33].

ANNs are based on the McCulloch-Pitts artificial neuron model. The model represents a neuron as a binary threshold unit andinputs are assigned weights before being summed, and compared against a specific threshold to determine the neuron's output. This effectively enables the representation of logical functions [36]. With the advent of backpropagation and activation functions -such as the Rectified Linear Unit (ReLU)- Deep Neural Networks (DNNs) further built upon the ANN model by increasing the number of hidden layers which enabled more complex patterns and representations to be modeled [37][38]. Deep learning models such as convolutional neural networks (CNNs), have more recently been applied to log file-based PSQA. CNNs are well-suited to image classification, making them ideal for use with fluence maps that can be generated by log file data. CNNs apply filters to detect desired features, reduce spatial dimensions to retain the most important features, and then perform final classification or predictions. They circumvent the need for manual feature selection. They are highly scalable for large datasets and have improved computational efficiency [39]. CNN's capabilities for detecting highly complex and time-dependent errors make them ideal for log file-based PSQA applications. They can identify small misalignments in MLC positions, irregular dose rate fluctuations, as well as other more subtle anomalies that may be missed by more traditional machine learning models. To prevent overfitting, large, labeled datasets are required and can be vulnerable to being misled by small input changes. CNN's decision making can be extremely difficult to interpret [40].

II. Examples of Recent Applications

This section will summarize the current machine learning applications for IMRT/VMAT PSQA within literature, including both drawbacks and advantages.

A. Recent Models for IMRT/VMAT PSQA

Most current applications for these models in IMRT and VMAT PSQA can be classified as either parameter prediction studies or error detection studies (see Table 1). Most parameter prediction studies are structured to predict GPR, with error detection studies predominantly performed on induced error data input. Either approach appears to struggle with similar limitations.

B. Drawbacks and Limitations

Tomori et al.[41], Lam et al.[43], Ono et al.[44], Huang et al.[45], Wang et al.[46], and Song et al.[47] used the parameter prediction approach. Using a prediction approach, all studies indicated that machine learning models could be effectively trained using log files to predict machine parameters at the time of treatment delivery for new treatment plans. These studies vary in the models explored, including SVM, RF, CNNs, and others. All models have relatively promising accuracy as seen in Table 1. However, Tomori et al.'s scope was limited to prostate IMRT plans, Huang et al. was limited to chest IMRT plans, and Song et al. was similarly limited to nasopharyngeal carcinoma and only used static gantry IMRT plans. Lam et al. included plans for multiple anatomical sites but were still specific to IMRT. Ono et al. and Wang et al. were specific to VMAT plans. Ono et al. and Lam et al. both performed their studies on multiple linear accelerators, but only Lam et al. used data from more than one institution. All six studies acknowledge that by using trajectory files, which are dependent on the linear accelerator itself, there is some vulnerability to machine-based error. As such, most log file-based PSQA is considered an enhancement to other QA measures that ensure the machine is calibrated appropriately, either with separate protocols or by incorporating additional sources of data into future models.

Error detection studies such as those by Kimura et al.[48], Sakai et al.[49], and Nyflot et al.[50] were similarly limited to one treatment plan type from a single institution. The only study that incorporated both VMAT and IMRT plans into a single study was an error detection study by Chuang et al. However, the study was only focused on MLC errors.

C. Positive Developments

These preliminary studies have gleaned significant insights into creating a holistic model for automating PSQA using log file data with a clear improvement upon methods over time. Lam et al. trained their model for predicting dosimetric effects in lieu of GPR to overcome any discrepancies between gamma index and errors that are clinically relevant [43]. Kimura et al. directly compared gamma map-based CNN models with dose difference map-based CNN models and found dose difference maps were more accurate [48]. Sakai et al. included radiomic data which resulted in higher sensitivity and specificity for MLC position and MLC modeling errors [49]. Hirashima et al. utilized a combination of 3D dosiomic features and plan complexity in a tree-based model [52]. Tomori et al.'s GPR prediction-based CNN model struggled with overestimating low GPR values and underestimating GPR in the test set [41][42]. Song et al. developed a novel model that weighed the MSE loss function to mitigate this class imbalance with promising results [47]. However, as all these studies have been limited to relatively small, single, or double institution datasets, their results are difficult to directly compare to one another. Additionally, most of the literature has been performed using Varian machines [21]. Although Varian machines are widely used in the US, Elekta machines are also used.

III. Discussion

Literature has broadly indicated that CNNs and other Deep Learning models appear to be the most successful at creating a model that is robust against certain biases seen in SVM and tree-based algorithms [53]. Although some studies have utilized data augmentation, most studies have agreed that to bring these findings to a clinically relevant standpoint, sufficient data must be collected from multiple institutions, techniques, treatment machines, and anatomical sites [54][55]. Additionally, encompassing both Varian and Elekta machines is essential to ensure this PSQA strategy is accurate on both platforms [56].

Furthermore, past work has predominantly focused on deterministic methods, which are ideal for providing direct, quantitative evaluations of dose delivery accuracy. While these are incredibly important in the overall application of the model, there are many aspects of treatment that carry uncertainty. Error tolerance, dose assessments, and multi-criteria evaluations are all subject to imprecision. Cilla et al. approached these aspects by using a "traffic light" protocol [57]. The protocol leveraged plan complexity to designate plans as acceptable (green light), requires further verifcation (orange light), or unacceptable (red light). Fuzzy logic follows similar reasoning and has been successfully applied to radiation control systems and treatment plan optimization [58][59]. Fuzzy logic uses fuzzy sets and linguistic variables to model uncertain or imprecise information. Desired variables can be assigned degrees of truth rather than a yes/no value. When applied to complex systems, this mathematical system eliminates the restriction of binary values to create more human-like decision making. The Fuzzy-CID3 (F-CID3) algorithm is a tree-based, hybrid method that combines neural networks and fuzzy sets, generating its own topology. Using a neural fuzzy number tree with a class separation method, the F-CID3 algorithm simplifies architecture compared to precesssors, achieving better performance with fewer connections [60].

IV. Conclusions and Future Directions

Recent work has proposed log file-based PSQA as a promising solution to the limitations of traditional phantom-based QA methods by leveraging Machine Learning algorithms to predict IMRT/VMAT QA outcomes and detect errors [6]. These algorithms, including SVMs, tree-based models, and CNNs, have demonstrated substantial progress in using log files for treatment plan verification.

While studies show the potential of log file-based PSQA, they also highlight key limitations. These include the inability to detect mechanical miscalibration or treatment planning errors, and the restricted scope of available data [45][47][61]. Mechanical errors can be mitigated through enhanced QA protocols for linacs and the incorporation of more sensitive machine tools. Moreover, issues such as insufficient training data for cancer site stratification and the lack of multi-institutional studies with diverse machine types remain significant barriers to widespread implementation [47][57][61][62][63].

Given the time-consuming nature of current PSQA protocols, log file-based PSQA, combined with AI model predictions, offers an efficient alternative. Future studies should focus on creating larger, multi-institutional datasets and exploring features within machine learning models that identify key factors in treatment failure. As machine learning and deep learning models evolve, their integration into clinical practice could lead to more efficient, accurate, and real-time quality assurance for radiation therapy.

References

[1] R. L. Siegel, A. N. Giaquinto, and A. Jemal, "Cancer statistics, 2024.", *CA: A Cancer Journal for Clinicians*, vol. 74, no. 1, pp. 12–49, Jan. 2024. DOI: 10.3322/caac.21820.

[2] A. A. for Cancer Research, *Aacr Cancer Progress Report 2024*. AACR, 2024, ISBN: 979-8-9857852-7-2.

[3] R. J. Hutten et al., "Worsening racial disparities in utilization of intensity modulated radiation therapy.", *Advances in radiation oncology*, vol. 7, no. 3, p. 100 887, Jan. 2022, ISSN: 24521094. DOI: 10.1016/j.adro.2021.100887.

[4] A. Agnew, C. E. Agnew, M. W. D. Grattan, A. R. Hounsell, and C. K. McGarry, "Monitoring daily MLC positional errors using trajectory log files and EPID measurements for IMRT and VMAT deliveries.", *Physics in Medicine and Biology*, vol. 59, no. 9, N49–63, May 2014. DOI: 10.1088/0031-9155/59/9/N49.

[5] C. E. Agnew, D. M. Irvine, and C. K. McGarry, "Correlation of phantom-based and log file patient-specific QA with complexity scores for VMAT.", *Journal of applied clinical medical physics / American College of Medical Physics*, vol. 15, no. 6, p. 4994, Nov. 2014. DOI: 10.1120/jacmp.v15i6.4994.

[6] N. Childress, Q. Chen, and Y. Rong, "Parallel/opposed: IMRT QA using treatment log files is superior to conventional measurement-based method.", *Journal of applied clinical medical physics / American College of Medical Physics*, vol. 16, no. 1, p. 5385, Jan. 2015. DOI: 10.1120/jacmp.v16i1.5385.

[7] B. Neal et al., "A clinically observed discrepancy between image-based and log-based MLC positions.", *Medical Physics*, vol. 43, no. 6, pp. 2933–2935, Jun. 2016. DOI: 10.1118/1.4949002.

[8] M. Barnes et al., "Insensitivity of machine log files to MLC leaf backlash and effect of MLC backlash on clinical dynamic MLC motion: An experimental investigation.", *Journal of applied clinical medical physics / American College of Medical Physics*, vol. 23, no. 9, e13660, Sep. 2022. DOI: 10.1002/acm2.13660.

[9] J. Waring, C. Lindvall, and R. Umeton, "Automated machine learning: Review of the state-of-the-art and opportunities for healthcare.", *Artificial Intelligence in Medicine*, vol. 104, p. 101 822, Apr. 2020, ISSN: 09333657. DOI: 10.1016/j.artmed.2020.101822.
[10] A. Esteva et al., "A guide to deep learning in healthcare.", *Nature Medicine*, vol. 25, no. 1, pp. 24–29, Jan. 2019, ISSN: 1078-8956. DOI: 10.1038/s41591-018-0316-z.
[11] K. Rasheed et al., "Explainable, trustworthy, and ethical machine learning for healthcare: A survey.", *Computers in biology and medicine*, vol. 149, p. 106 043, Oct. 2022. DOI: 10.1016/j.compbiomed.2022.106043.
[12] K. Y. Ngiam and I. W. Khor, "Big data and machine learning algorithms for health-care delivery.", *The Lancet Oncology*, vol. 20, no. 5, e262–e273, May 2019. DOI: 10.1016/S1470-2045(19)30149-4.
[13] J. M. Park, J.-I. Kim, and H.-G. Wu, "Technological advances in charged-particle therapy.", *Cancer research and treatment : official journal of Korean Cancer Association*, vol. 53, no. 3, pp. 635–640, Jul. 2021. DOI: 10.4143/crt.2021.706.
[14] R. A. Chandra, F. K. Keane, F. E. M. Voncken, and C. R. Thomas, "Contemporary radiotherapy: Present and future.", *The Lancet*, vol. 398, no. 10295, pp. 171–184, Jul. 2021. DOI: 10.1016/S0140-6736(21)00233-6.
[15] J. Bertholet, Y. Vinogradskiy, Y. Hu, and D. J. Carlson, "Advances in image-guided adaptive radiation therapy.", *International Journal of Radiation Oncology, Biology, Physics*, vol. 110, no. 3, pp. 625–628, Jul. 2021. DOI: 10.1016/j.ijrobp.2021.02.047.
[16] S. Bartzsch et al., "Technical advances in x-ray microbeam radiation therapy.", *Physics in Medicine and Biology*, vol. 65, no. 2, 02TR01, Jan. 2020. DOI: 10.1088/1361-6560/ab5507.
[17] C. X. Yu, C. J. Amies, and M. Svatos, "Planning and delivery of intensity-modulated radiation therapy.", *Medical Physics*, vol. 35, no. 12, pp. 5233–5241, Dec. 2008. DOI: 10.1118/1.3002305.
[18] A. F. I. Osman, N. M. Maalej, and K. Jayesh, "Prediction of the individual multileaf collimator positional deviations during dynamic IMRT delivery priori with artificial neural network.", *Medical Physics*, vol. 47, no. 4, pp. 1421–1430, Apr. 2020. DOI: 10.1002/mp.14014.
[19] P. Szeverinski et al., "Error sensitivity of a log file analysis tool compared with a helical diode array dosimeter for VMAT delivery quality assurance.", *Journal of applied clinical medical physics / American College of Medical Physics*, vol. 21, no. 11, pp. 163–171, Nov. 2020. DOI: 10.1002/acm2.13051.
[20] E. E. Klein et al., "Task group 142 report: Quality assurance of medical accelerators.", *Medical Physics*, vol. 36, no. 9, pp. 4197–4212, Sep. 2009. DOI: 10.1118/1.3190392.
[21] G. A. Ezzell et al., "IMRT commissioning: Multiple institution planning and dosimetry comparisons, a report from AAPM task group 119.", *Medical Physics*, vol. 36, no. 11, pp. 5359–5373, Nov. 2009. DOI: 10.1118/1.3238104.
[22] M. Miften et al., "Tolerance limits and methodologies for IMRT measurement-based verification QA: Recommendations of AAPM task group no. 218.", *Medical Physics*, vol. 45, no. 4, e53–e83, Apr. 2018. DOI: 10.1002/mp.12810.
[23] R. A. C. Siochi, A. Molineu, and C. G. Orton, "Point/counterpoint. patient-specific QA for IMRT should be performed using software rather than hardware methods.", *Medical Physics*, vol. 40, no. 7, p. 070 601, Jul. 2013. DOI: 10.1118/1.4794929.
[24] J. C. Smith, S. Dieterich, and C. G. Orton, "Point/counterpoint. it is still necessary to validate each individual IMRT treatment plan with dosimetric measurements before delivery.", *Medical Physics*, vol. 38, no. 2, pp. 553–555, Feb. 2011. DOI: 10.1118/1.3512801.
[25] D. W. Litzenberg, J. M. Moran, and B. A. Fraass, "Verification of dynamic and segmental IMRT delivery by dynamic log file analysis.", *Journal of applied clinical medical physics / American College of Medical Physics*, vol. 3, no. 2, pp. 63–72, 2002. DOI: 10.1120/jacmp.v3i2.2578.
[26] A. L. McNiven, M. B. Sharpe, and T. G. Purdie, "A new metric for assessing IMRT modulation complexity and plan deliverability.", *Medical Physics*, vol. 37, no. 2, pp. 505–515, Feb. 2010. DOI: 10.1118/1.3276775.
[27] J. M. Moran et al., "Safety considerations for IMRT: Executive summary.", *Practical radiation oncology*, vol. 1, no. 3, pp. 190–195, Sep. 2011. DOI: 10.1016/j.prro.2011.04.008.
[28] L. Masi, R. Doro, V. Favuzza, S. Cipressi, and L. Livi, "Impact of plan parameters on the dosimetric accuracy of volumetric modulated arc therapy.", *Medical Physics*, vol. 40, no. 7, p. 071 718, Jul. 2013. DOI: 10.1118/1.4810969.
[29] B. Sun et al., "Initial experience with TrueBeam trajectory log files for radiation therapy delivery verification.", *Practical radiation oncology*, vol. 3, no. 4, e199–208, Dec. 2013. DOI: 10.1016/j.prro.2012.11.013.
[30] C. Cortes and V. Vapnik, "Support-vector networks", *Machine learning*, vol. 20, no. 3, pp. 273–297, Sep. 1995, ISSN: 0885-6125. DOI: 10.1007/{BF00994018}.
[31] T. K. Ho, "Random decision forests", in *Proceedings of 3rd International Conference on Document Analysis and Recognition*, vol. 1, 1995, 278–282 vol.1. DOI: 10.1109/ICDAR.1995.598994.
[32] J. H. Friedman, "Greedy function approximation: A gradient boosting machine", *The Annals of Statistics*, vol. 29, no. 5, pp. 1189–1232, Oct. 2001, ISSN: 0090-5364. DOI: 10.1214/aos/1013203451.
[33] T. Chen and C. Guestrin, "XGBoost: A scalable tree boosting system", in *Proceedings of the 22nd ACM SIGKDD International Conference on Knowledge Discovery and Data Mining - KDD '16*, New York, New York, USA: ACM Press, Aug. 2016, pp. 785–794, ISBN: 9781450342322. DOI: 10.1145/2939672.2939785.
[34] R. Tibshirani, "Regression shrinkage and selection via the lasso", *Journal of the Royal Statistical Society: Series B (Methodological)*, vol. 58, no. 1, pp. 267–288, Jan. 1996, ISSN: 00359246. DOI: 10.1111/j.2517-6161.1996.tb02080.x.
[35] A. E. Hoerl and R. W. Kennard, "Ridge regression: Biased estimation for nonorthogonal problems", *Technometrics : a journal of statistics for the physical, chemical, and engineering sciences*, vol. 12, no. 1, pp. 55–67, Feb. 1970, ISSN: 0040-1706. DOI: 10.1080/00401706.1970.10488634.
[36] W. S. McCulloch and W. Pitts, "A logical calculus of the ideas immanent in nervous activity", *The Bulletin of mathematical biophysics*, vol. 5, no. 4, pp. 115–133, Dec. 1943, ISSN: 0007-4985. DOI: 10.1007/{BF02478259}.
[37] A. G. Ivakhnenko, "The group method of data handling; a rival of the method of stochastic approximation", 1968.
[38] R. H. Hahnloser, R. Sarpeshkar, M. A. Mahowald, R. J. Douglas, and H. S. Seung, "Digital selection and analogue amplification coexist in a cortex-inspired silicon circuit.", *Nature*, vol. 405, no. 6789, pp. 947–951, Jun. 2000. DOI: 10.1038/35016072.
[39] Y. Lecun, L. Bottou, Y. Bengio, and P. Haffner, "Gradient-based learning applied to document recognition", *Proceedings of the IEEE*, vol. 86, no. 11, pp. 2278–2324, 1998. DOI: 10.1109/5.726791.
[40] W. Ding, M. Abdel-Basset, H. Hawash, and A. M. Ali, "Explainability of artificial intelligence methods, applications and challenges: A comprehensive survey", *Information sciences*, vol. 615, pp. 238–292, Nov. 2022, ISSN: 00200255. DOI: 10.1016/j.ins.2022.10.013.

[41] S. Tomori et al., "A deep learning-based prediction model for gamma evaluation in patient-specific quality assurance.", *Medical Physics*, Jul. 2018. DOI: 10.1002/mp.13112.

[42] S. Tomori et al., "Systematic method for a deep learning-based prediction model for gamma evaluation in patient-specific quality assurance of volumetric modulated arc therapy.", *Medical Physics*, vol. 48, no. 3, pp. 1003–1018, Mar. 2021. DOI: 10.1002/mp.14682.

[43] D. Lam et al., "Predicting gamma passing rates for portal dosimetry-based IMRT QA using machine learning.", *Medical Physics*, vol. 46, no. 10, pp. 4666–4675, Oct. 2019. DOI: 10.1002/mp.13752.

[44] T. Ono et al., "Prediction of dosimetric accuracy for VMAT plans using plan complexity parameters via machine learning.", *Medical Physics*, vol. 46, no. 9, pp. 3823–3832, Sep. 2019. DOI: 10.1002/mp.13669.

[45] Y. Huang et al., "Deep learning for patient-specific quality assurance: Predicting gamma passing rates for IMRT based on delivery fluence informed by log files.", *Technology in Cancer Research & Treatment*, vol. 21, p. 15 330 338 221 104 881, 2022. DOI: 10.1177/15330338221104881.

[46] L. Wang et al., "Multi-task autoencoder based classification-regression model for patient-specific VMAT QA.", *Physics in Medicine and Biology*, vol. 65, no. 23, p. 235 023, Nov. 2020. DOI: 10.1088/1361-6560/abb31c.

[47] W. Song et al., "Improving the performance of deep learning models in predicting and classifying gamma passing rates with discriminative features and a class balancing technique: A retrospective cohort study.", *Radiation Oncology*, vol. 19, no. 1, p. 98, Jul. 2024. DOI: 10.1186/s13014-024-02496-5.

[48] Y. Kimura, N. Kadoya, S. Tomori, Y. Oku, and K. Jingu, "Error detection using a convolutional neural network with dose difference maps in patient-specific quality assurance for volumetric modulated arc therapy.", *Physica medica : PM : an international journal devoted to the applications of physics to medicine and biology : official journal of the Italian Association of Biomedical Physics (AIFB)*, vol. 73, pp. 57–64, May 2020. DOI: 10.1016/j.ejmp.2020.03.022.

[49] M. Sakai et al., "Detecting MLC modeling errors using radiomics-based machine learning in patient-specific QA with an EPID for intensity-modulated radiation therapy.", *Medical Physics*, vol. 48, no. 3, pp. 991–1002, Mar. 2021. DOI: 10.1002/mp.14699.

[50] M. J. Nyflot, P. Thammasorn, L. S. Wootton, E. C. Ford, and W. A. Chaovalitwongse, "Deep learning for patient-specific quality assurance: Identifying errors in radiotherapy delivery by radiomic analysis of gamma images with convolutional neural networks.", *Medical Physics*, vol. 46, no. 2, pp. 456–464, Feb. 2019. DOI: 10.1002/mp.13338.

[51] K.-C. Chuang, W. Giles, and J. Adamson, "A tool for patient-specific prediction of delivery discrepancies in machine parameters using trajectory log files.", *Medical Physics*, vol. 48, no. 3, pp. 978–990, Mar. 2021. DOI: 10.1002/mp.14670.

[52] H. Hirashima et al., "Improvement of prediction and classification performance for gamma passing rate by using plan complexity and dosiomics features.", *Radiotherapy and Oncology*, vol. 153, pp. 250–257, Dec. 2020, ISSN: 01678140. DOI: 10.1016/j.radonc.2020.07.031.

[53] A. F. I. Osman and N. M. Maalej, "Applications of machine and deep learning to patient-specific IMRT/VMAT quality assurance.", *Journal of applied clinical medical physics / American College of Medical Physics*, vol. 22, no. 9, pp. 20–36, Sep. 2021. DOI: 10.1002/acm2.13375.

[54] Y. Interian et al., "Deep nets vs expert designed features in medical physics: An IMRT QA case study.", *Medical Physics*, vol. 45, no. 6, pp. 2672–2680, Jun. 2018. DOI: 10.1002/mp.12890.

[55] D. M. Fondevila, P. J. Rios, D. M. D. Peñalva, and S. Arbiser, "Predicting gamma passing rates for portal dosimetry-based IMRT QA using deep learning.", *International Journal of Radiation Oncology, Biology, Physics*, vol. 111, no. 3S, e110–e111, Nov. 2021. DOI: 10.1016/j.ijrobp.2021.07.515.

[56] I. J. Das et al., "Accelerator beam data commissioning equipment and procedures: Report of the TG-106 of the therapy physics committee of the AAPM.", *Medical Physics*, vol. 35, no. 9, pp. 4186–4215, Sep. 2008. DOI: 10.1118/1.2969070.

[57] S. Cilla et al., "Prediction and classification of VMAT dosimetric accuracy using plan complexity and log-files analysis.", *Physica medica : PM : an international journal devoted to the applications of physics to medicine and biology : official journal of the Italian Association of Biomedical Physics (AIFB)*, vol. 103, pp. 76–88, Nov. 2022. DOI: 10.1016/j.ejmp.2022.10.004.

[58] T.-F. Lee et al., "A fuzzy system for evaluating radiation treatment plans of head and neck cancer", in *2012 9th International Conference on Fuzzy Systems and Knowledge Discovery*, 2012, pp. 510–514. DOI: 10.1109/FSKD.2012.6234043.

[59] C. Pinter, T. Olding, L. J. Schreiner, and G. Fichtinger, "Using fuzzy logics to determine optimal oversampling factor for voxelizing 3D surfaces in radiation therapy", *Soft computing*, vol. 24, no. 24, pp. 18 959–18 970, Dec. 2020, ISSN: 1432-7643. DOI: 10.1007/s00500-020-05126-w.

[60] L. Sztandera, "Computational intelligence foundations", in *Computational Intelligence in Business Analytics: Concepts, Methods, and Tools for Big Data Applications*, 1st ed., Upper Saddle River, New Jersey: Pearson, 2014, ISBN: 0-13-355208-X.

[61] Y. Huang et al., "A feasibility study to predict 3D dose delivery accuracy for IMRT using DenseNet with log files.", *Journal of X-ray science and technology*, vol. 32, no. 4, pp. 1199–1208, 2024. DOI: 10.3233/{XST}-230412.

[62] P. Viola et al., "Prediction of VMAT delivery accuracy using plan modulation complexity score and log-files analysis.", *Biomedical physics & engineering express*, vol. 8, no. 5, Aug. 2022. DOI: 10.1088/2057-1976/ac82c6.

[63] T. Ono et al., "Applications of artificial intelligence for machine- and patient-specific quality assurance in radiation therapy: Current status and future directions.", *Journal of radiation research*, vol. 65, no. 4, pp. 421–432, Jul. 2024. DOI: 10.1093/jrr/rrae033.

[64] J. N. K. Carlson et al., "A machine learning approach to the accurate prediction of multi-leaf collimator positional errors.", *Physics in Medicine and Biology*, vol. 61, no. 6, pp. 2514–2531, Mar. 2016. DOI: 10.1088/0031-9155/61/6/2514.

[65] D. A. Granville, J. G. Sutherland, J. G. Belec, and D. J. La Russa, "Predicting VMAT patient-specific QA results using a support vector classifier trained on treatment plan characteristics and linac QC metrics.", *Physics in Medicine and Biology*, vol. 64, no. 9, p. 095 017, Apr. 2019. DOI: 10.1088/1361-6560/ab142e.

[66] C. Ma et al., "The structural similarity index for IMRT quality assurance: Radiomics-based error classification.", *Medical Physics*, vol. 48, no. 1, pp. 80–93, Jan. 2021. DOI: 10.1002/mp.14559.

[67] P. D. Wall and J. D. Fontenot, "Application and comparison of machine learning models for predicting quality assurance outcomes in radiation therapy treatment planning", *Informatics in Medicine Unlocked*, vol. 18, p. 100 292, 2020, ISSN: 23529148. DOI: 10.1016/j.imu.2020.100292.

[68] K. S. Lew et al., "Prediction of portal dosimetry quality assurance results using log files-derived errors and machine learning techniques.", *Frontiers in oncology*, vol. 12, p. 1 096 838, 2022. DOI: 10.3389/fonc.2022.1096838.

TABLE I. SUMMARY OF RECENT STUDIES USING MACHINE LEARNING MODELS FOR IMRT/VMAT PSQA. (AUC= AREA UNDER THE CURVE, MAE= MEAN ABSOLUTE ERROR, RMSE= ROOT MEAN SQUARE ERROR, SR= SPEARMAN'S RANK CORRELATION COEFFICIENT)

Author/Year	Plan Type	Dataset Size	Anatomic Sites	Algorithm	QA Outcome	Feature Count	Key Results
Carlson et al. 2016 [64]	VMAT	74 plans (3,161,280 data points)	Multiple	RF	Error detection	6	RMSE= 0.193mm (linear regression)
Tomori et al. 2018 [41]	IMRT	60 plans	Prostate	CNN	Parameter prediction	N/A	Errors within 1.10% at 3%/3mm criteria
Interian et al. 2018 [54]	IMRT	498 plans	Multiple	CNN	Parameter prediction	N/A	MAE= 0.70% at 3%/3mm criteria
Lam et al. 2019 [43]	IMRT	1497 beams	Multiple	Tree-based	Parameter prediction	31	Errors within 3% for 98% of predictions at 2%/2mm criteria
Ono et al. 2019 [44]	VMAT	600 plans	Multiple	Regression Tree, ANN, Other	Parameter prediction	28	Mean prediction error= -0.2% at 3%/3mm criteria (ANN)
Granville et al. 2019 [65]	VMAT	1,620 beams	Multiple	SVM	Error detection	60	AUC=0.88 (macro-averaged)
Nyflot et al. 2019 [50]	IMRT	186 beams (558 images)	Multiple	SVM, Decision Tree, Other	Error detection	145	Accuracy= 64.3% for SVM
Ma et al. 2020 [66]	IMRT	180 beams (1,620 images)	Multiple	SVM, RF, Other	Error detection	276	AUC=0.86 for linear SVM
Osman et al. 2020 [18]	IMRT	10 plans (360,800 datapoints)	Multiple	ANN	Error detection	14	RMSE=0.0096mm
Wall and Fontenot 2020 [67]	VMAT	500 plans	Multiple	SVM, Tree-Based, ANN	Parameter prediction	241	MAE=3.75% at 3%/3mm criteria (SVM)
Hirashima et al. 2020 [52]	VMAT	1,255 plans	Multiple	Tree-based	Parameter prediction	875	MAE=4.2% and AUC=0.83 at 2%/2mm criteria
Wang et al. 2020 [46]	VMAT	276 Plans	Multiple	ANN	Parameter prediction	N/A	Absolute prediction error=1.76% at 3%/3mm criteria
Kimura et al. 2020 [48]	VMAT	161 Beams	Prostate	CNN	Error detection	54	Accuracy=0.94
Tomori et al. 2020 [42]	VMAT	147 plans	Multiple	CNN	Parameter prediction	N/A	MAE=0.63% at 3%/3mm criteria
Sakai et al. 2021 [49]	IMRT	38 beams (152 error plans)	Multiple	SVM, Tree-based, Other	Error detection	837	AUC=1.00 for leaf transmission factor error, 1.0 for dosimetric leaf gap error, 0.80 for leaf positional error vs. error free (SVM)
Chuang et al. 2021 [51]	IMRT/VMAT	267 IMRT and VMAT plans (10,584,120 data points)	Multiple	Tree-based, Other	Error detection	7	RMSE=0.0085 mm (Boosted Tree Model)
Huang et al. 2022 [45]	IMRT	112 plans	Chest	CNN	Parameter prediction	4	MAE and RMSE decreased with stricter gamma criteria, while SR and R^2 increased as gamma criteria were made stricter (3%/3mm, 3%/2mm, 2%/3mm, and 2%/2mm)
Cilla et al. 2022 [57]	VMAT	651 plans/1,302 arcs	Multiple	SVM, Other	Parameter prediction	3	Precision of 93.1 for gamma % and 92.7% for gamma mean for the testing dataset at 2%/2mm (SVM)
Lew et al. 2022 [68]	VMAT	578 log files	Multiple	RF, SVM, Other	Parameter prediction	13	Average error of less than 2% with 1%/1mm criteria.
Song et al. 2024 [47]	IMRT	204 plans/2,348 fields	Nasopharyngeal Carcinoma	CNN	Parameter prediction	1-8	AUC= 0.92 with 0.77 sensitivity and 0.89 specificity

Data Mining Techniques in Online Health Communities

Cassandra Mikkelson and Cali Sweitzer
College of Life Sciences
Thomas Jefferson University
Philadelphia, Pennsylvania
e-mail: {cassandra.mikkelson | cali.sweitzer}@students.jefferson.edu

***Abstract*—Online health communities are an untapped domain of unlimited data on patient sentiment towards drugs and medical devices that can provide academia and industry an inside scope of in demand research according to patient responses. These communities are often found on social media platforms, such as Facebook and Reddit, where patients who have similar medical histories connect to share their experiences, advice, and support for each other. This review explores how data mining methods, specifically machine learning and Natural Language Processing (NLP), can be applied to analyze large data sets derived from user-generated responses on social media and health databases. Methods discussed include sentiment analysis, clustering algorithms, and text classification models as effective tools to generate new knowledge on patterns within online health discussions. The paper also highlights potential applications of data mining to improve pharmaceutical research, enhance drug monitoring, and identify adverse events in terms of post-market surveillance for regulatory bodies like the U.S Food and Drug Administration (FDA). Lastly, challenges related to data transformation, cleaning, and privacy concerns are addressed along with proposed augmentations to improve data quality.**

Keywords-data mining; online health communities; Patient Sentiment; Sentiment Analysis; Healthcare Data Transformation.

I. Introduction

In the age of social media, online communities have become a haven where patients facing health challenges can exchange insights and share common experiences. Online health communities are formed on social media platforms including Facebook groups and subreddits on Reddit, where patients and caregivers come together to share their experiences, advice, and support for others within their communities. Patient-driven platforms including Patientslikeme, Health Union, and Healthboards are networks specifically formulated for user camaraderie in the healthcare setting, unlike those support groups naturally formed on other social medias. These platforms and online health communities empower patients to act as healthcare consultants, in the form of reviewing drugs, devices, surgeries, and specific healthcare providers [1]. Consequently, these communities generate an abundance of self-reported data on patient sentiment, which provides valuable insight into patient satisfaction connected to health services.

Data mining combines machine learning, algorithms, statistical analysis, artificial intelligence, and database management systems [2]. Once a database of interest is defined, the data is transformed to complement the model that is created. The model is then tested, evaluated, and interpreted to generate new knowledge through the generation of a custom report. Data mining enables the user to analyze data across different dimensions that recapitulate useful information that can inspire new ideas.

Data mining techniques can serve as potential tools to extract previously hidden information and patterns from online communities. A machine learning approach that most effectively scans, processes, and summarizes social media data would be NLP, which includes text classification (e.g., Support Vector Machines (SVM) and naïve Bayes classifier), sentiment analysis, clustering algorithms (e.g., Self-Organizing Map (SOM)), and supervised learning algorithm (e.g., decision tree).

The results generated by mining of large patient-derived datasets could inform the pharmaceutical industry about in demand medical interventions according to patient needs. Consequently, research outputs could positively impact current pre-clinical and clinical trials to streamline desired research according to patient sentiment and break the barrier between the bedside and benchtop.

This review article highlights various data mining techniques that could be utilized to collect and transform data from online health communities. Section I introduces the concepts of online health communities and data mining. Section II proposes different data mining methods that could help reduce the complexity of data obtained from these communities, including sentiment analysis, SOM, SVM, and the naïve Bayes classifier. Section III explores how data collected from such studies could be applied by the U.S. Food and Drug Administration (FDA) to identify adverse drug reactions and improve efficiency in drug production, such as vaccine development. Section IV addresses challenges related to data transformation and cleaning, proposing augmentations like web scraping and the Levenshtein distance method to address issues associated with data collection from online forums. Finally, Section V concludes the article, emphasizing the underutilization of data generated by online health communities and its potential to positively influence academic and institutional medical research.

II. Proposed Methods and Techniques

As patients increasingly turn to online communities and health platforms for reviews, advice, and solidarity, the development of mining techniques to analyze this user-generated data has become more essential. Alnashwan et al. described three data-driven approaches to elucidate patient sentiment within these online forums: sentiment analysis, content analysis, and topic analysis. Sentiment analysis is a broad field of study with the goal of identifying and characterizing the emotional tone of a body of text [3]. Sentiment analysis is widely used in the context of understanding patient values, attitudes, and preferences towards medical providers, prescriptions and treatments, and adverse effects. It serves as a classification model within supervised machine learning, where predictions are made and validated through associated characteristics. Classic sentiment analysis groups mined posts based on three categories: positive, negative, and neutral. Some studies in current literature also calculated the degree of emotion using a numerical scale, e.g., -5 to +5, based on keywords within the text. Alnashwan et al. hypothesized that classifying medical posts on a binary (positive/negative) or polarity (degree of sentiment) based scale would not be sufficient to encompass the broad and complex nature of online health-related text [3]. As such, the authors suggest a bottom-up categorization approach, in which posts are manually mined for specific sentiment-based keywords and subsequently grouped into seed categories. The multitude of seed categories is then further filtered into six core categories based on the predominating sentiment, examples including treatment inquiry, symptom confusion, and seeking general information. Once categorized, different techniques of data mining under the umbrella of sentiment analysis can be employed. Such techniques include the use of machine learning and lexicon-based text classification systems.

One of many examples of machine learning techniques includes the use of Microsoft's Azure Machine Learning software, which serves as a resource for data scientists and machine learning engineers to generate programs such as NLP using artificial intelligence. These tools are developed through machine learning in which the user builds algorithms that allow the computer to continually learn based on predictive models [4]. Such models including NLP where the computer becomes able to interpret and categorize informal text are crucial for data mining of patient sentiment in online health communities.

This broad concept of NLP and sentiment analysis includes the use of lexicon-based text classification systems: content analysis and topic analysis. Content analysis is a research method used to extract meaningful content within a large body or dataset of text by analyzing and grouping relationships of high frequency words, phrases, or themes [4]. In conjunction with content analysis, this NLP technique can also employ topic analysis, which aims to identify overarching topics within a body of text based on a probabilistic model [4].

Simultaneously, Jawad et al. proposed two techniques for text classification that can be used together to identify patterns in patient sentiment from social media in an article for the Proceedings of the 2017 Future Technologies Conference [5]. A SOM categorizes input vectors based on a wordlist into a neural network, hence creating clusters based off words defined as positive and negative. This model would utilize the Term-Frequency-Inverse Document Frequency (TF-IDF) to vectorize text files by assigning text with a numerical statistic that would interpret the frequency of a word in a document relative to the whole document. TF reflects the frequency of a given word and IDF reflects the rarity of a word. The implementation of this model would output results that categorize responses in terms of a positive or negative sentiment, which would be useful to pharmaceutical companies to gain knowledge on patient assessment of their products.

Alternatively, techniques used in data mining from mobile health apps would be translational to mining for online health communities. Fallah et al. compiled a systemic review on the common data mining methods correlated with health apps. They found that the top three successful methods with the highest level of accuracy were cloud-based SVM, decision tree, and naïve Bayesian [2]. After data has been vectorized, the SVM classifies the data on a binary scale and generates a separating hyperplane line that separates the two groups. This method would best be used for a yes/no research question (e.g., is the response to a product positive or negative?). The decision tree mimics a tree, with population classified in branches that construct a tree with roots, internal nodes, and leaf nodes. Nodes reflect choices made in a decision that splits into a branch that represents the outcome of the decision. Data is split into parent nodes and child nodes, to decide the category of the text file. It is commonly used for creating classifications based on a prediction algorithm. Meanwhile, the naïve Bayes classifier vectorizes the text files into multi-dimensional numerical probability values that are used as input for the SOM, SVM, or decision tree for the final classification step. Probability values are based on the probability of a text that contains pre-defined words is equal to the probability of finding these pre-defined words in a category [6].

Through data mining methods, the complexity and breadth of public online data are reduced to expose undiscovered patterns in common patient sentiment and reveal previously unreported ailments. Researchers must design their model according to how their question would categorize their data. For example, SVM is most appropriate for binary classification, while SOM is best for multiple categories and would complement a study to discover the adverse effects of a product from patient reviews.

III. APPLICATIONS

The applications of data mining in the healthcare setting are robust, allowing for the harnessing of vast amounts of relevant online medical data. One application of data mining in healthcare includes the identification of adverse drug reactions. Some relevant adverse reactions are not apparent until after clinical trial testing and approval by the FDA, as factors which may cause these adverse reactions are often difficult to account for in the clinical trial period. After approval by the FDA, important factors that may cause adverse reactions include long-term use, co-exposures to other drugs, environmental and dietary variances, as well as genetic differences that may not have been probable to account for during clinical trials [7].As such, databases such as the FDA Adverse Event Reporting System (FAERS) give critical insights into relevant drug reactions. As outlined in the most recent FDA White Paper on Data Mining from 2018, several techniques are applied to FAERS safety reports to explicate possible adverse events [8]. Disproportionality methods are used to identify statistically significant associations between medications and events. One such method includes the use of the Proportional Reporting Ratio (PRR) in which the degree of reporting of an adverse event for a particular drug is compared to the same event occurrence amongst all reports of all drugs within the FAERS database [8]. Thus, this robust data serves as a baseline for the occurrence of any event, allowing for associations to be made based on disproportionate reporting. Statistical methods beyond this data mining technique are then employed to further validate causative rather than correlative relationships [8].

Another clinically relevant database used by researchers and data miners alike to extrapolate patient-derived data includes the FDA's Vaccine Adverse Event Reporting System (VAERS). Data mining for patient sentiment in the context of vaccine efficacy and reactogenicity has become increasingly relevant following the COVID-19 pandemic. As described by Dror et al., vaccination compliance relies on a personal risk-benefit perception which can be skewed by misinformation and perceived side effects that may not align with scientific evidence [9]. Data mining and subsequent analysis of such reports proves as an effective tool for minimizing misinformation regarding vaccine reactogenicity, potentially enhancing vaccine uptake [10]. Data mining of VAERS reports provides a powerful dataset for understanding public sentiment related to vaccines, with direct relevancy to vaccine uptake.

IV. Challenges and Proposed Augmentations

Generating knowledge from a larger data set is generally a challenging and time-consuming task. This is especially so when the data in social media communities contains about two decades of responses, including spelling errors and abbreviations that would make creating a word list an ambitious effort. Transforming and cleaning this unorganized data would take an extended amount of time as these proposed models are best suited for survey responses. To minimize this task, web scraping can be used to extract data from unstructured web browsers into structured data that can be used for analysis. Web scraping is the process of data transformation through computer software, that mimics human behaviors of web exploration to compile data more efficiently than by hand [11]. Meanwhile, there is still the struggle to correct spelling errors and abbreviations, which may be resolved by utilizing Levenshtein distance to identify errors by comparison against a dictionary.

Patient sentiment may vary according to several demographic and clinical characteristics, particularly the social determinants of health—non-medical factors like race, ethnicity, religion, and socioeconomic status—that significantly influence a person's health [12]. As previously mentioned, data mining can be used to extract previously unknown information hidden within a data set. Thus, data mining has the power to segment data according to the social determinants of health and could uncover the diversity in patient sentiment according to factors defined by the social determinants of health. However, extracting such personal data would be challenging and might require social media profile information. To address this, researchers could create tools to categorize responses according to these factors before applying data mining techniques. Web scraping software could also aid in this task.

Studies that result from this work would have to address ethical concerns of informed consent, since consent cannot be obtained from all social media platform users. Even though studies will be compiled of public data from online forums, the collection of data must be aligned with data privacy laws, which protects users from the collection of their names to comply with confidentiality and anonymity. Before the conduction of a study on data mining from online health communities, researchers must ensure compliance with federal and institutional laws and policies along with the privacy policies of the social media platform of interest.

V. Conclusion and Future Work

The goal of this review article is to summarize how data mining can be a useful tool to collect information from online health communities. With the power of data mining, valuable results can be obtained to steer the current pharmaceutical field in the direction of patient-centered research to drive drug and device development toward medical interventions desired by the patient. These methods are currently involved in mining databases, like FAERS and VAERS, for information on adverse drug related events and vaccine uptake. Therefore, the suggested tools are relevant and applicable to be translated to online health communities. Potential techniques were proposed along with the challenges that will be faced and suggested augmentations that will require future work. Further research would include the implementation of these methods and techniques to generate a report from health data obtained from online communities. As a result of these endeavors, upcoming biomedical research would be fueled by patient-centered data.

References

[1] M.-G. Fayn, V. des Garets, and A. Rivière, "Collective empowerment of an online patient community: Conceptualizing process dynamics using a multi-method qualitative approach," *BMC Health Services Research*, vol. 21, pp. 1–19, 2021.

[2] M. Fallah and S. R. N. Kalhori, "Systematic review of data mining applications in patient-centered mobile-based information systems," *Healthcare informatics research*, vol. 23, no. 4, pp. 262–270, 2017.

[3] R. Alnashwan, A. O'Riordan, and H. Sorensen, "Multiple-perspective data-driven analysis of online health communities," in *Healthcare*, MDPI, vol. 11, 2023, p. 2723.

[4] Y. Mao, Q. Liu, and Y. Zhang, "Sentiment analysis methods, applications, and challenges: A systematic literature review," *Journal of King Saud University-Computer and Information Sciences*, p. 102 048, 2024.

[5] M. S. Jawad, W. Adi, A. Salem, and M. Doiher, "Implementation of data mining from social media for improved public health care," *Future Technologies Conference*, pp. 234–240, 2017.

[6] D. Isa, V. Kallimani, and L. H. Lee, "Using the self organizing map for clustering of text documents," *Expert Systems with Applications*, vol. 36, no. 5, pp. 9584–9591, 2009.

[7] A. Bone and K. Houck, "The benefits of data mining," *Elife*, vol. 6, e30280, 2017.

[8] U. Food and D. Administration, *Data mining at FDA*, U.S. Food and Drug Administration website, 2018.

[9] A. A. Dror *et al.*, "Vaccine hesitancy: The next challenge in the fight against covid-19," *European journal of epidemiology*, vol. 35, no. 8, pp. 775–779, 2020.

[10] M. D. Rousculp *et al.*, "Burden and impact of reactogenicity among adults receiving covid-19 vaccines in the United states and Canada: Results from a prospective observational study," *Vaccines*, vol. 12, no. 1, p. 83, 2024.

[11] M. A. Khder, "Web scraping or web crawling: State of art, techniques, approaches and application.," *International Journal of Advances in Soft Computing & Its Applications*, vol. 13, no. 3, pp. 144–168, 2020.

[12] M. Marmot and R. Wilkinson, *Social determinants of health*. Oxford Academic, 2005.

Detecting Suicide Risk and Exploring Contributing Factors: Classification and Topic Modeling of Social Media Data

Evan Dan, Jianfeng Zhu, Ruoming Jin
Department of Computer Science,
Kent State University
Kent, USA
e-mail: edan1@kent.edu, jzhu10@kent.edu, rjin1@kent.edu

Abstract—**Suicide remains a critical global health issue, with over 700,000 lives lost annually. Existing research has explored factors influencing suicidal thoughts, but traditional studies often rely on small-scale data sources that may overlook contextual influences. This study aims to address that gap by analyzing a large dataset of posts from Reddit communities r/SuicideWatch and r/Teenagers to detect suicidal ideation and identify associated themes. Using Natural Language Processing and statistical methodologies, including Llama 3-8b and Mistral-7b, we fine-tuned models with manually labeled data to improve classification accuracy of posts for suicidal ideation. Using data re-labeled by the large language models, BERTopic identified key themes linked to suicidal ideation: relationship struggles, academic stress, and family trauma. While non-suicidal posts also included social and academic concerns, the topics were centered around more immediate stressors rather than the long-term emotional distress issues seen in the suicidal group. These findings highlight the potential of NLP methodologies in analyzing large-scale social media data, offering valuable insights for informing new prevention strategies. Additionally, social media, in combination with NLP, serves as a valuable outlet for capturing genuine emotional struggles, enabling more timely and personalized mental health support compared to traditional approaches like counseling.**

Keywords-Suicide; The Llama 3-8b; Mistral-7b; GPT-4o; Reddit; BERTopic Modeling; contributing factors.

I. INTRODUCTION

Each year, 726,000 people around the world lose their lives to suicide, with many more attempting it daily [1]. Youth are particularly vulnerable, with suicide being the second leading cause of death for those in aged 10–14 and 25–34 years and the third leading cause for individuals aged 15–24 years in the U.S. in 2022 [2][3]. However, beyond impacting the conflicted individuals, suicide also leaves lasting effects on these individuals' families and communities. A survey conducted by Cerel et al on 1,736 adults in Kentucky found that suicide-exposed individuals, those personally affected by suicide, were twice as likely as unexposed individuals to meet screening criteria, which assess mental health symptoms, for depression and nearly twice as likely for anxiety [4]. Additionally, suicide is driven by various social, cultural, biological, psychological, and environmental factors that span a lifetime. By examining child suicide cases from the National Violent Death Reporting System, Ruch et al identified four key themes: mental health and suicide concerns (31.4%), traumatic experiences (27.1%), family challenges (39.8%), and school or peer difficulties (35.6%) [5]. Building on this, Turecki et al emphasized the interplay of genetics, personality traits, psychiatric illnesses, and environmental influences in suicide risk [6].

Although these studies offer important insights, conventional tools such as questionnaires and surveys present challenges in thoroughly uncover and examine the complexities of suicidal factors. Due to limitations like structured response formats, inherent biases, and the absence of dynamic, real-time data, these methods frequently fall short in capturing the nuanced feelings and experiences of individuals with suicidal ideation [7]. Furthermore, research based on suicide reporting databases often focus on deaths by suicide, excluding data on non-fatal suicide attempts or suicidal ideation, which are critical for understanding the full spectrum of suicidal behavior. However, studies have demonstrated that non-fatal suicide attempts are significant predictors of future suicide risk, with research by Turecki et al indicating that past suicide attempts increase the likelihood of subsequent suicidal behavior [6]. In other words, these data sources often lack the depth and scope needed to fully understand the range of suicidal behaviors, including nuanced emotions and non-fatal attempts, which are critical for assessing future suicide risks.

On the other hand, as social media, especially Reddit, play a growing role in mental health discussions, they have emerged as promising data sources due to its role in fostering open discussions. Reddit, with over 97 million daily active users [8], features mental health subreddits like r/SuicideWatch (512K+ members), where users can share feelings and seek help [9]. Its anonymity and diverse subcommunities encourage users to discuss sensitive topics more openly, making it a unique platform for gathering authentic mental health data, as highlighted by Yeskuatov et al [10]. This was exemplified during the COVID-19 pandemic, when Reddit saw spikes in posts about health anxiety, economic stress, social isolation, and substance use [11][12][13], providing valuable insights into mental health trends. These studies underscore the platform's potential in providing crisis support and fostering connection during challenging times.

Building on the potential of innovative data sources like Reddit, recent research has also highlighted the importance of advanced analytical approaches to better understand and predict suicidal behaviors. For instance, Franklin et al.'s meta-analysis called for the use of complex Machine Learning (ML) models to enhance predictive accuracy in identifying suicidal thoughts and behaviors [14]. Advancements in ML and Natural Language Processing (NLP) offer promising tools for analyzing mental health data and identifying patterns related to suicidal behaviors [15], facilitating deeper exploration of mental health discussions on platforms like Reddit. For instance, Bauer et al utilized large language models (LLMs) to analyze Reddit posts, revealing patterns of disconnection, hopelessness, and trauma in users experiencing suicidality [16]. Expanding on these approaches, BERTopic, developed by Grootendorst, offers several advantages over traditional clustering regression techniques [17]. BERTopic has been used to detect signs of depression on Reddit, analyze public sentiment towards artificial intelligence (AI) in mental health, and track mental health trends during the COVID-19 pandemic [18][19][20], demonstrating its potential for advancing mental health research.

Traditional research on suicide risk has largely depended on standardized survey methods or government datasets, which often fall short in representing the emotional complexity of individuals struggling with suicidal thoughts. These methods tend to focus narrowly on isolated factors, overlooking the intricate and interconnected nature of suicidal behaviors. In contrast, our study leverages a substantial dataset of Reddit posts from r/SuicideWatch and r/Teenagers, enabling access to vast, real-time, and unfiltered expressions of emotional states. By utilizing advanced LLMs, such as Llama 3-8b and Mistral-7b, coupled with sophisticated topic modeling techniques, we were able to identify nuanced factors associated with suicidal thoughts, classify posts as suicidal or non-suicidal with precision, and uncover detailed themes within these categories. This approach overcomes the limitations of traditional datasets, providing a deeper, more comprehensive understanding of suicide-related behavior. The use of large language models allows us to capture intricate patterns, emotional nuances, and contextual insights that are otherwise inaccessible through conventional methods. By expanding the scope of analysis and enhancing its depth, our findings provide actionable, evidence-based strategies to inform suicide prevention efforts and foster meaningful advancements in the field.

The main contributions of this paper can be summarized as follows:

1. Demonstrated value of using Natural Language Processing methodologies, including fine-tuning Llama 3-8b and Mistral-7b, for analyzing social media data regarding suicidal ideation, a topic full of complex nuances.
2. The Llama 3-8b model achieved a test accuracy of 0.9371 for classifying Reddit posts for suicidal ideation, demonstrating its ability to capture detailed emotional patterns.
3. Using BERTopic, we revealed key topics in the discussions within the classified suicidal and non-suicidal Reddit posts.

The rest of this paper is organized as follows. Section II provides a thorough description of the data preprocessing and analysis methods, detailing the cleaning process, classification models, evaluation metrics, and topic modeling approaches. Section III presents and discusses the main findings. Section IV finishes the paper with the conclusions.

II. METHODS

This next section will detail the specific steps taken within the study to reach the analysis results obtained.

A. Data Collection and Preprocessing

This study utilized a Kaggle dataset compiling posts from December 16, 2008, to January 2, 2021, sourced from the subreddits r/SuicideWatch and r/Teenagers. The dataset uploader, Komati, anonymized the usernames for pseudonymity [21]. The data was then preprocessed to improve analysis accuracy. We employed the Pandas library to streamline analysis by arranging the unprocessed data into structured data frames [22]. Regular Expressions (RegEx) was applied to remove repetitive filler text, unnecessary whitespace (newlines, spaces, tabs), and URLs, while also converting all text to lowercase to maintain consistency within the data [23]. Similarly, we utilized the Unidecode library to remove accented characters [24]. In addition, contractions were converted to their complete forms using the Contractions library [25] while lemmatization, the process of transforming words to their base forms, was applied using the NLTK library [26].

In the original dataset, posts were categorized as "suicide" or "non-suicide" based on the subreddit they originated from [21]. However, there were inaccuracies with this labeling, so we re-classified the posts to improve the accuracy of the labels using large LLMs. We first manually labeled approximately 900 posts as "suicidal" or "non-suicidal" according to the definition of "suicidal ideation" from the Diagnostic and Statistical Manual of Mental Disorders [27]. Afterwards, the hand-labeled data was divided into three partitions: 80% of the data for training, 10% of validation, and 10% for testing. Two LLMs, Llama 3-8b and Mistral 7-b, were trained on this data, and the more accurate model was selected to re-label the full dataset.

B. Detection and Classification Models

Llama 3-8B is an advanced LLM developed by Meta and was released in April of 2024. Consisting of 8 billion parameters, it can interpret and classify textual data effectively. It was pretrained on 15 trillion tokens of publicly available data, with the data having gone through Llama 2 models and advanced data-filtering pipelines to ensure their high quality [28]. Additionally, with methods such as supervised fine-tuning (SFT) and reinforcement learning with human feedback (RLHF), Meta was able to ensure the model provides accurate responses while prioritizing safety [29]. Because of the extensive pretraining, Llama 3-8B is well-suited for identifying complex patterns and nuanced meanings in textual data, making it an ideal tool for classifying Reddit posts.

Released in October of 2023, Mistral 7B is an advanced LLM that was developed by Mistral AI [30]. The model contains 7.3 billion parameters and is designed for both speed and precision for Natural Language Processing (NLP) tasks. The model incorporates innovative mechanisms such as Grouped-Query Attention (GQA) and Sliding Window Attention (SWA). GQA is an optimization technique that reduces computational complexity and enhances the efficiency of predictions [31]. Meanwhile, SWA enables the model to process longer text inputs at a manageable computational cost. Demonstrated in Figure 1, it achieves this by employing a sliding window that restricts the model's focus to a small segment of the input at a time. By processing many of these windows individually and sliding them across the input sequence, the model can utilize its multiple transformer blocks to identify indirect relationships between tokens across these segments. This approach ensures it can effectively comprehend the contexts of long inputs while maintaining computational efficiency [30]. With these methods, Mistral 7B is well-suited for classifying the Reddit posts as features such as the SWA make it ideal for capturing nuanced patterns often present in lengthy posts.

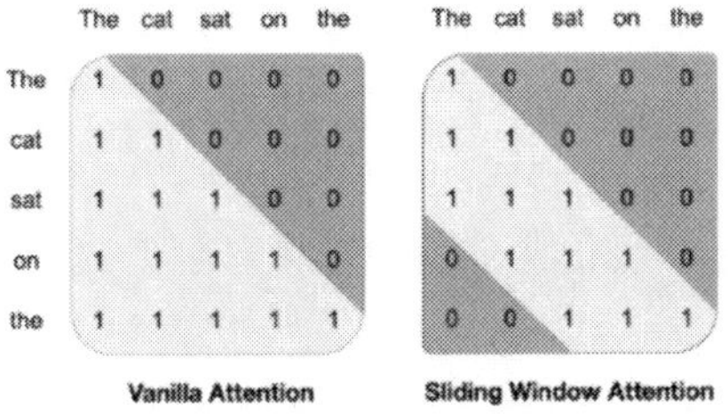

Figure 1. Sample Demonstration of SWA [30].

The Llama 3-8b and Mistral-7b models were each trained on manually labeled posts for three epochs with a learning rate of 0.0002. To assess the effectiveness of the final model, we employed standard evaluation metrics, including accuracy, precision, recall, and F-1 scores. These metrics are defined using four key terms that are defined below.

- **True Positive (TP):** Model correctly identifies a suicidal post as suicidal.
- **True Negative (TN):** Model correctly identifies a non-suicidal post as non-suicidal.
- **False Positive (FP):** Model incorrectly identifies a non-suicidal post as suicidal (Type 1 Error).
- **False Negative (FN):** Model incorrectly identifies a suicidal post as non-suicidal (Type 2 Error).

Accuracy measures the overall proportion of correct predictions with a ratio of all correctly classified posts to the total number of posts [32]:

$$\text{Accuracy} = \frac{TP + TN}{TP + FP + TN + FN} \tag{1}$$

Precision measures the proportion of predicted positive cases that are correct [32] with a ratio of correctly classified suicidal posts to all posts predicted as suicidal (including both correctly and incorrectly identified suicidal posts):

$$\text{Precision} = \frac{TP}{TP + FP} \tag{2}$$

Recall assesses the portion of actual positive cases that are identified correctly [32] with a ratio of correctly classified suicidal posts to all posts from r/SuicideWatch (including both correctly identified suicidal posts and incorrectly identified non-suicidal posts):

$$\text{Recall} = \frac{TP}{TP + FN} \tag{3}$$

In situations where there is a high cost associated with false-negative prediction; recall proves to be very useful for identifying the best model [33]. Between false positives and false negatives, it is most likely less consequential to incorrectly predict someone as suicidal than to incorrectly predict a suicidal person as non-suicidal.

F1-Score is the harmonic mean of precision and recall [32]. In other words, the F1-Score is an average of the two metrics:

$$\text{F1} = 2 \times \frac{Precision \times Recall}{Precision + Recall} \tag{4}$$

To evaluate the status of the models during the fine-tuning process, we utilized training loss and validation loss scores. After completing the fine-tuning, we selected the versions of the Llama 3-8B and Mistral 7B models with the lowest losses to test against the test dataset to ensure optimal performance. The model with the higher test accuracy was then used to classify the rest of data for suicidal ideation. Finally, the processed and re-labeled dataset was separated into two categories based on the suicide/non-suicide label. This step enables targeted analysis focused on addressing the research objectives of understanding factors associated with suicide, as this allows us to analyze and compare results between the two groups.

C. *BERTopic*

To analyze the re-classified data, BERTopic was employed for topic modeling to uncover key themes within the posts. The model incorporates BERT embeddings, improving its capacity to detect subtle patterns and capture nuanced sentiments within the dataset [17]. Additionally, by utilizing contextual term frequency-inverse document frequency (c-TF-IDF), BERTopic ensures precise clustering of topics with meaningful and coherent groupings. The analysis was applied to each group of classified posts, "suicidal" and "non-suicidal", facilitating a thorough comparison of their characteristics and providing deeper insights into the themes specific to each group. In addition, we included the KeyBERTInspired model [34] to filter out stopwords, improving the topics' precision of the clusters. We also incorporated OpenAI's GPT-4o [35] to generate specific and detailed topic labels, enhancing the clarity and interpretability of the identified themes. The following prompt was used to guide the labeling process:

I have a topic that contains the following documents: [DOCUMENTS]

The topic is described by the following keywords: [KEYWORDS].

On the basis of the information above, extract a short but highly descriptive topic label. Make sure it is in the following format: topic: <topic label>.

III. Results

Table I showcases examples of posts before and after processing, highlighting the impact of the cleaning process. In the cleaned versions, the texts exhibit a more uniform and consistent structure while retaining the original sentiments of the posts.

TABLE I. SAMPLE OF DATA PREPROCESSING

Class	Text before and after pre-processing	
	Original Post	***Cleaned Text***
non-suicide	Finally 2020 is almost over... So I can never hear "2020 has been a bad year" ever again. I swear to fucking God it's so annoying	finally 2020 is almost over.. so i can never hear "2020 has been a bad year" ever again. i swear to fucking god it is so annoying
non-suicide	i need help just help me im crying so hard	i need help just help me i am crying so hard
suicide	It ends tonight.I can't do it anymore. \nI quit.	it end tonight.i can not do it anymore. i quit.
suicide	Been arrested - feeling suicidal Edit	been arrested - feeling suicidal edit

A. Model Performance

Model performance for Llama 3-8B and Mistral-7B model can be seen in Table II below. With slightly a slightly higher accuracy, recall, and F1 score, the Llama 3-8B model outperformed the Mistral-7B against the test dataset, so the Llama 3-8B model was selected to re-label the remaining data. Afterwards, the model re-labeled 96,086 posts as "suicidal" and 135,988 posts as "non-suicidal".

TABLE II. MODEL PERFORMANCE COMPARISON

Model	Test Performance			
	Accuracy	***Recall***	***Precision***	***F1 Score***
Llama 3-8b	0.9371	0.9371	1.000	0.9676
Mistral-7b	0.9314	0.9314	1.000	0.9645

B. Key Topics Identified by BERTopic

To address the goal of uncovering themes linked to suicidal thoughts, this study employed BERTopic to analyze both suicidal and non-suicidal posts. For analysis, we concentrated on specific topics of concerns to align with the study's aim of investigating the underlying factors associated with suicidal thoughts.

By focusing on these themes, the analysis maintains a strong commitment to the research objectives. Table III highlights the top 10 topics of concern identified for the suicidal group while Table IV displays the top 10 topics of concern for the non-suicidal group.

Expanding on the top 10 topics identified by BERTopic in Table III, three critical themes emerged related to sources of mental instability across various life stages: relationship struggles affecting adults' emotional health, academic stress impacting students' self-esteem, and family trauma influencing childhood development. The most common theme in the suicidal group was centered around struggles and emotional turmoil from relationships (3,285 posts), suggesting that difficulties in maintaining or coping with personal relationships may contribute to feelings of hopelessness often associated with suicidal ideation. The emotional attachment and dependence involved in romantic relationships can lead to profound loneliness or a loss of self-identity when disrupted, prompting some to contemplate suicide. Afterwards, academic pressures emerged as the second most common topic (1,791 posts), illustrating how the fear of failure can contribute to feelings of anxiety and hopelessness, especially among students. This finding may reflect the increased academic demands and societal expectations placed on students, which can result in feelings of inadequacy. Another prominent theme, the childhood and family trauma topic (1,421 posts) underlines the lasting impacts of early life experiences on mental health. Such traumas can contribute to complex emotional issues and unresolved feelings that may intensify over time, particularly as individuals face adulthood. When compounded by present-day struggles, these lingering effects can potentially trigger suicidal thoughts.

TABLE III. TOP 10 TOPICS IN SUICIDAL POSTS

Rank	Topics	
	Suicide Topic (n=39742)	***Count***
1	Experiencing emotional struggles and breakups in relationships	3285
2	Struggles with academic failure and mental health	1791
3	Childhood and family trauma	1421
4	Emotional dilemmas with suicide	1273
5	Suicidal intent with overdosing on pills	878
6	Struggles with loneliness and low self-esteem	856
7	Suicidal farewell messages	787
8	Self-harm and suicidal ideation	538
9	Depressed birthdays and suicidal thoughts	392
10	Suicidal intents with firearms	347

Moreover, themes such as "Emotional Dilemmas with Suicide" (1,273 posts) and "Suicidal Intent with Overdosing on Pills" (878 posts) reveal that some users are not only coping with underlying struggles but are also directly confronting the act of suicide itself. These users express deep internal conflicts about their thoughts or contemplate specific methods, reflecting the internal turmoil between the desire to escape emotional pain and the emotional, moral, or religious beliefs that deter them from acting on suicidal impulses.

Additionally, the topics of loneliness and low self-esteem (856 posts) and self-harm (538 posts) suggest that personal insecurities, such as feelings of isolation and worthlessness, may exacerbate mental health struggles and contribute to suicidal ideation. These themes may point to specific psychological patterns that may serve as focal points for early intervention and tailored support strategies, as feelings of worthlessness can contribute to fragile mental states, while self-harm behaviors can escalate into the risk of suicidal thoughts. Moreover, the "Suicidal Farewell Messages" topic (787 posts) and the "Suicidal Intents with

Firearms" topic (347 posts) highlight further expressions of extreme distress experienced by some individuals. Given the lethality of firearms and the commitment reflected in discussions of suicide notes, these topics may suggest progression from ideation to preparation, revealing the importance of identifying these specific signals to prevent further progression through timely intervention.

TABLE IV. TOP 10 TOPICS IN NON-SUICIDAL POSTS

Rank	Topics	
	Non-suicide Topic (n=63885)	*Count*
1	Exploring and managing dynamics in romantic relationships	6711
2	Concerns with boredom and loneliness	3286
3	Academic struggles for college students	1813
4	Struggles with parents and family dynamics	1121
5	Sexual frustrations	611
6	Challenges in socializing for introverts	536
7	Challenges with sleeping and persistent insomnia	440
8	Issues related to racism and discriminatory language	431
9	Concerns with alcohol consumption	407
10	Struggles of transgender identity	393

By comparison, while non-suicidal posts in Table IV reflect significant challenges, they do not indicate immediate crises but reflect more general concerns of teenagers and young adults. Comparing these two sets of topics reveals some thematic overlaps, including topics regarding romantic relationship struggles, academic pressures, and family dynamics, which are experienced and expressed in distinct ways by each group. In the suicidal posts, these themes are often associated with feelings of hopelessness, personal failure, or a desire to escape, whereas, in non-suicidal posts, the same issues appear to provoke irritation, uncertainty, or a desire for improvement. For instance, the topic regarding relationship struggles for suicidal posts is more centered toward long-term emotional distress, whereas the corresponding topic for non-suicidal posts reflects a focus on exploring emotional dynamics and personal growth within relationships. Similarly, the academic topic in the non-suicidal group is geared more towards daily struggles, whereas the topic for the suicidal group reveals deeper feelings of perceived failure accompanied by intense self-criticism.

Broader social and personal concerns such as discussions about racism and transgender identities highlight personal challenges in managing mental and emotional well-being, while topics such as boredom and loneliness demonstrate feelings of disconnection and a lack of purpose. While these topics do not explicitly indicate suicidal ideation, they may represent underlying problems that could escalate into more severe issues more, such as low self-esteem, which are more commonly observed in suicidal posts. In contrast, themes regarding farewell messages and detailed suicide planning point to a deeper level of emotional distress, reflecting a serious stage of suicidal intent.

This comparison indicates the importance of understanding context and intensity within mental health discussions. Interventions for suicidal individuals should prioritize crisis management and emotional support tailored to severe psychological distress. For non-suicidal individuals, interventions might instead focus on counseling, life skills training, and support networks that help them manage common stressors before they become more severe.

IV. CONCLUSIONS

This study demonstrates the effectiveness of advanced NLP and statistical techniques in identifying and analyzing suicidal ideation based on large-scale social media data from Reddit. By fine-tuning Llama 3-8B, Mistral 7B, and BERTopic, we revealed key sources of mental instability associated with suicidal thoughts, including relationship struggles, academic stress, and family trauma. The findings also highlighted distinct thematic differences between posts indicating suicidal ideation and general adolescent concerns, revealing deeper insights into specific triggers and expressions of suicidal thoughts among young individuals. Our results underscore the potential of NLP in real-time mental health monitoring and intervention on social media. Fine-tuned models, such as Llama 3-8B, which achieved a test accuracy of 0.9371, demonstrate strong predictive performance, offering scalable tools for distress detection. By addressing warning signs before they escalate into crises, these systems can provide early interventions in teenage communities, leveraging the thematic overlaps between suicidal and non-suicidal groups to design broader mental health support initiatives. To enhance model accuracy, future research could incorporate data from a wider range of social media platforms to improve generalizability while also exploring the influence of digital interactions on users' mental health. Expanding this work will offer deeper insights into the evolving role of social media in mental health and help develop targeted intervention strategies tailored to specific online behaviors and mental health challenges.

REFERENCES

[1] World Health Organization. Suicide. Published August 23, 2023, Available: https://www.who.int/news-room/fact-sheets/detail/suicide, [retrieved: Feburary, 2025].

[2] Centers for Disease Control and Prevention. Web-based Injury Statistics Query and Reporting System (WISQARS): leading causes of death reports 2001-2020, national and regional, 2020, Available: https://webappa.cdc.gov/sasweb/ncipc/leadcaus10_us.html, [retrieved: Feburary, 2025].

[3] CDC. Mental Health Is A Growing Problem. Published 2022, Available:https://www.cdc.gov/healthyyouth/mental-health/index.htm, [retrieved: Feburary, 2025].

[4] J. Cerel, et al. "Exposure to Suicide in the Community: Prevalence and Correlates in One U.S. State." Public Health Rep. vol.131(1), pp.100-107, 2016, doi:10.1177/003335491613100116, https://pmc.ncbi.nlm.nih.gov/articles/PMC4716477/

[5] D. A. Ruch, et al. "Characteristics and Precipitating Circumstances of Suicide Among Children Aged 5 to 11

Years in the United States, 2013-2017." JAMA Netw Open. Vol. 4(7), e2115683 2021, doi:10.1001/jamanetworkopen.2021.15683.

[6] G. Turecki, et al. "Suicide and suicide risk." Nat Rev Dis Primers. vol.5(1):74, 2019. doi:10.1038/s41572-019-0121-0.

[7] A. Gulliver, K. M. Griffiths and H. Christensen, "Perceived barriers and facilitators to mental health help-seeking in young people: a systematic review." BMC Psychiatry.10:113, 2010. doi:10.1186/1471-244X-10-113.

[8] Reddit, Reddit by the Numbers, Available: https://www.redditinc.com/press, [retrieved: Feburary, 2025].

[9] Reddit, Peer Support for Anyone Struggling with Suicidal Thoughts, Available: https://www.reddit.com/r/SuicideWatch/ [retrieved: Feburary, 2025].

[10] E. Yeskuatov, S. L. Chua and L. K. Foo. "Leveraging Reddit for suicidal ideation detection: a review of machine learning and natural language processing techniques." Int J Environ Res Public Health. vol.19(16), 10347 2022, Doi:10.3390/ijerph191610347.

[11] D. M. Low, et al. "Natural language processing reveals vulnerable mental health support groups and heightened health anxiety on reddit during COVID-19: observational study." J Med Internet Res. vol.22(10), e22635, 2020, doi:10.2196/22635.

[12] J. Zhu, et al. "Investigating COVID-19's Impact on Mental Health: Trend and Thematic Analysis of Reddit Users' Discourse." J Med Internet Res. vol. 25, e46867, 2023, doi:10.2196/46867.

[13] C. McAuliffe, et al. "Connectedness in the time of COVID-19: Reddit as a source of support for coping with suicidal thinking." SSM Qual Res Health. vol.2, 100062, 2022; doi:10.1016/j.ssmqr.2022.100062.

[14] J. C. Franklin, et al. "Risk Factors for Suicidal Thoughts and Behaviors: A Meta-Analysis of 50 Years of Research." Psychol Bull., vol. 143, pp. 187–232, 2017 doi: 10.1037/bul0000084.

[15] A. L. Glaz, et al. "Machine Learning and natural language processing in mental health: systematic review." J Med Internet Res. vol.23(5), e15708, 2021. doi:10.2196/15708.

[16] B. Bauer, et al. "Using Large Language Models to Understand Suicidality in a Social Media–Based Taxonomy of Mental Health Disorders: Linguistic Analysis of Reddit Posts." JMIR Ment Health vol. 11, e57234, 2024, DOI: 10.2196/57234, https://mental.jmir.org/2024/1/e57234

[17] M. Grootendorst, "BERTopic: neural topic modeling with a class-based TF-IDF procedure." arXiv. Published March 11, 2022. doi:10.48550/arXiv.2203.05794.

[18] P. D. Thushari, et al. "Identifying discernible indications of psychological well-being using ML: explainable AI in reddit social media interactions." Soc Netw Anal Min. vol. 13(141), 2023. doi:10.1007/s13278-023-01145-1.

[19] A. Dhankar and A. Katz. "Tracking pregnant women's mental health through social media: an analysis of reddit posts", JAMIA Open. Vol.6(4), 2023,
doi:10.1093/jamiaopen/ooad094.

[20] Y. Cai, F. Wang, H. Wang and Q. Qian, "Public sentiment analysis and topic modeling regarding ChatGPT in mental health on Reddit: negative sentiments increase over time." arXiv. Published November 27, 2023.
doi:10.48550/arXiv.2311.15800.

[21] Kaggle, Suicide and Depression Detection, Available: https://www.kaggle.com/nikhileswarkomati/suicide-watch, [retrieved: Feburary, 2025].

[22] Pandas. PyPI, Published January 30, 2020, Available: https://pypi.org/project/pandas/, [retrieved: Feburary, 2025].

[23] M. Barnett, regex: Alternative regular expression module, to replace re. PyPI, Available: https://pypi.org/project/regex/, [retrieved: Feburary, 2025].

[24] T. Solc, Unidecode: ASCII transliterations of Unicode text. PyPI, Available: https://pypi.org/project/Unidecode/, [retrieved: Feburary, 2025].

[25] P. V. Kooten,. contractions: Fixes contractions such as `you're` to you `are`. PyPI, Available: https://pypi.org/project/contractions/, [retrieved: Feburary, 2025].

[26] S. Bird, et al. nltk: Natural Language Toolkit. PyPI, Available: https://pypi.org/project/nltk/, [retrieved: Feburary, 2025].

[27] Amercican Psychiatric Association, "Diagnostic and Statistical Manual of Mental Disorders", 5th ed. American Psychiatric Association, 2013.

[28] Meta, Introducing Meta Llama 3: The most capable openly available LLM to date, Published April 2024, Available: https://ai.meta.com/blog/meta-llama-3/, [retrieved: Feburary, 2025].

[29] Meta AI, Meta Llama 3 - 8B, Hugging Face, Available: https://huggingface.co/meta-llama/Meta-Llama-3-8B, [retrieved: Feburary, 2025].

[30] P. Khadka, Mistral 7B explained, Published May 2024, Available: https://medium.com/@pranjalkhadka/mistral-7b-explained-53720dceb81e, [retrieved: Feburary, 2025].

[31] DataCamp, Mistral 7B Tutorial: A Step-by-Step Guide to Using and Fine-Tuning Mistral 7B, Available: https://www.datacamp.com/tutorial/mistral-7b-tutorial, [retrieved: Feburary, 2025].

[32] Basu, T. and Murthy, C. "A feature selection method for improved document classification." Advanced Data Mining and Applications: 8th International Conference, ADMA 2012, Nanjing, China, December 15-18, 2012. Proceedings 8. Springer Berlin Heidelberg, 2012.

[33] Evidently AI, Accuracy vs. Precision vs. Recall in Machine Learning: What's the Difference? Evidently AI, Available: https://www.evidentlyai.com/classification-metrics/accuracy-precision-recall. [retrieved: Feburary, 2025]..

[34] M. Grootendorst, KeyBERT API, KeyBERT, Available: https://maartengr.github.io/KeyBERT/api/keybert.html, [retrieved: Feburary, 2025].

[35] OpenAI, API Reference, platform.openai.com, Available: https://platform.openai.com/docs/api-reference/chat, [retrieved: Feburary, 2025].

BARRIER: Beta-Secretase 1 Reduction for Amyloid Plaque Regulation through Inhibition Exploration and Research

Neel Banga
Dougherty Valley High School
San Ramon, CA, United States
neelbanga21@gmail.com

Abstract—**Alzheimer's is a brain disorder that disproportionately affects older adults with its primary symptom being severe dementia. Worldwide, over 55 million people have Alzheimer's, with 6.7 million affected individuals living in the USA. Current methods to mitigate the effects of Alzheimer's are insufficient with most drugs (e.g., Memantine, Donepezil, Rivastigmine, etc.) being inconsistent while also causing heavy side effects. In order to address these issues, more drugs need to be tested for viability. To speed up the process, this research proposes AI-based models that can potentially detect which drugs will be able to effectively inhibit the crux of the Alzheimer's pathway, an enzyme named Beta Secretase 1. This study documented the investigation of four AI models—K-Nearest Neighbors (KNN), Random Forest, ChemBERTa, and PubChem10M—and their ability to predict drug efficacy for inhibiting BACE1, a vital target in the Alzheimer's Disease (AD) pathway. These models were trained on the ChEMBL4822 database. The KNN and RandomForest models were traditional descriptor-based models whereas the ChemBERTa and PubChem10M models were fine-tuned transformers. The KNN model showed a strong training performance of (R^2 = 0.6092); this score stayed consistent in the testing phase (R^2 = 0.6210). While having a lower score, the RandomForest model displayed similar consistency in the training (R^2 = 0.5651) and testing phase (R^2 = 0.5605). The ChemBERTa model showed significant improvement from the training phase (R^2 = 0.2641) to the testing one (R^2 = 0.6433), indicating high generalization potential. Similarly, the PubChem10M model exhibited large growth from the training (R^2 = 0.2641) to the testing phase (R^2 = 0.6194). These results highlight the unique strengths of each model and underscore the promising role of AI in AD drug discovery. Future work on the refinement and integration of these models could lead to more effective therapeutic agents for AD.**

Keywords-alzheimer's; beta-secretase 1; machine learning; transformer model; drug discovery

I. Introduction

Alzheimer's Disease (AD) is a neurodegenerative brain disorder, a type of brain disorder where cells in the central nervous system either fail to work or exist at all [1]. AD has debilitating symptoms (see Figure 1). In the USA alone, nearly 7 million individuals suffer from ***Alzheimer's***; this number is projected to rise to 13 million by 2050. Worldwide, ***Alzheimer's*** and similar dementia are presumed to affect over 55 million individuals and this number does not seem to be going away anytime soon [2].

Currently, most drugs in the market are unable to inhibit the progression of AD, rather they aim to cope with the effects that come with AD (donepezil [3], rivastigmine [4], memantine [5], etc.). The drugs that are able to inhibit the pathway are often controversial, expensive, and come with heavy side effects like brain swelling and microhemorrhages (memantine, lecanemab [6], etc.). With an unfortunate assortment of drugs that aren't able to completely eradicate the disease nor its effects, it is pivotal to find a drug that can effectively inhibit the spread of AD.

It is known that an overexposure/overproduction of Amyloid Plaques in the brain is synonymous with AD [7]; symptoms such as memory loss, poor judgment, lack of spontaneity, reduced cognitive ability, etc. occur when a plaque buildup is formed. Amyloid plaques are abnormal deposits of amino acid chains known as beta-amyloid peptides (Aβ). These are caused by the incorrect cleavage of the Amyloid Precursor Protein (a type 1 transmembrane protein), powered by ***Beta-Secretase*** 1 (BACE1) [8]—see Figure 2.

It is predicted that ***machine learning*** models trained on molecular descriptors and protein structural features will effectively predict IC50 scores for candidate drugs, aiding in identifying compounds with high efficacy in inhibiting ***Beta-Secretase 1*** (BACE1). This predictive capability directly influences the progression of ***Alzheimer's*** disease by enabling the discovery of potent inhibitors targeting the formation of Beta-Amyloid Peptides.

Amyloid plaques do not paint the complete story, however. Tau, an abundant protein in nerve cells, gives neurodegenerative properties to AD [9]. In a healthy organism (without AD), Tau proteins are primarily responsible for stabilizing microtubules. Tau binds to microtubules, ensuring their stability. It assists in nutrient transport within neurons and plays a role in cell division. Aβ and tau interact early in AD pathogenesis, even before the formation of plaques and tangles. Aβ modulates protein kinases and phosphatases, so an overproduction leads to tau misfolding and hyperphosphorylation. Neurofibrillary tangles form within neurons. These tangles consist of aggregated and hyperphosphorylated tau proteins. The accumulation of neurofibrillary tangles disrupts normal neuronal function. Tau tangles b lock communication between neurons, altering memory, cognition, and other brain functions. Tau-induced damage occurs at the synaptic level, where synapses (connections between neurons) are lost. This contributes to cognitive decline in AD. Acetylcholine, a neurotransmitter that plays a vital role in memory, learning, etc [1], is often unable to reach the brain in the presence of a toxic Tau protein therefore causing an acetylcholine deficiency in the brain and propagates the effects of Alzheimer's. Toxic tau enhances Aβ toxicity via

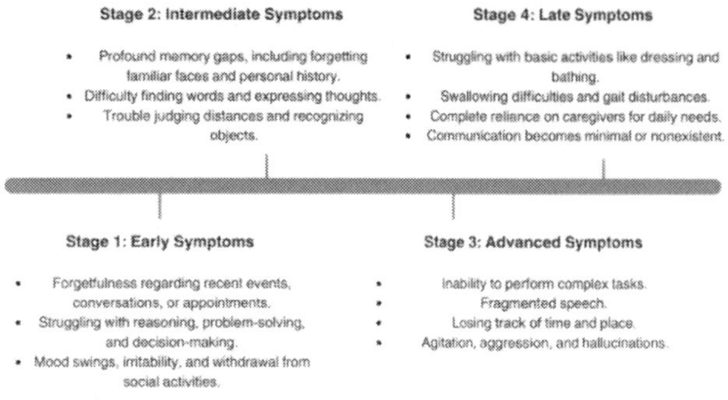

Figure 1. ***Alzheimer's*** Disease Symptoms

a feedback loop, therefore enhancing the symptoms of AD. This leads to a self-propagation of Tau and Aβ (see Figure 2) [10].

Inhibition of BACE1 would cause the suppression of AD due to the absence of Aβ and therefore prevent the formation of a toxic Tau protein. BACE1 is a critical target for the AD pathway as it has an early role in Beta-Amyloid protein production, and produces mild phenotype reactions when deleted, suggesting that inhibiting this enzyme might not have severe side effects and has a well-documented history due to aspartic protease identity. In recent years, there has been a surge in the use of Artificial Intelligence (AI) technology in the medical field. [11]. Various models and architectures have been utilized in biomedical research to enhance its scope and effectiveness. One prominent model that has garnered a lot of attention in recent years is the ***transformer model***. The idea is based on an attention mechanism: a mechanism that allows computers to weigh and understand the context behind different words [12]. This type of model is extremely diverse; it can be used in classification tasks, generative tasks, and even regression tasks [13]. This research does not only focus on the use of transformers; It is important to evaluate multiple models as different models work best for different use cases.

This research is no exception to the use of AI: the goal is to create deep learning models to predict whether drugs can inhibit BACE1 and subsequently find drugs that can disrupt ***Alzheimer's***. For this reason, this project will likely result in an AI model that can accurately identify drugs to inhibit BACE1, as well as find drugs that show large promise to suppress ***Alzheimer's*** [14]. This study explores AI models, including transformers, to predict drugs capable of inhibiting BACE1 and, by extension, tackling Alzheimer's while recognizing that IC50 values are part of a more comprehensive evaluation of drug efficacy. The aim is to develop AI models to identify potential drugs that can effectively inhibit BACE1 and explore promising candidates to mitigate AD progression.

In the related work and methods section, we discuss the related work and methodologies that underpin this study. The methodology section details the datasets and preprocessing techniques employed to prepare our data for analysis. In results,

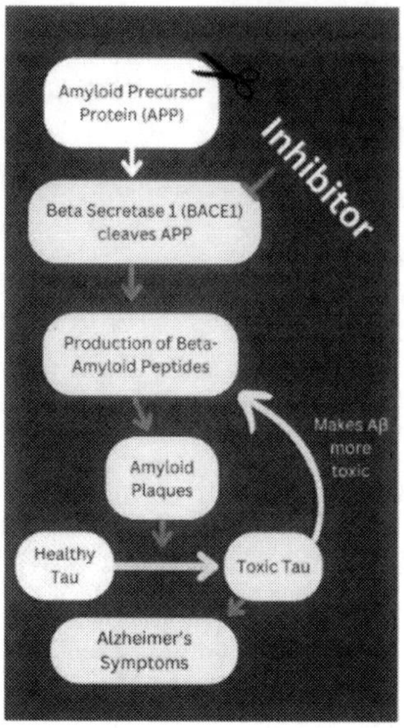

Figure 2. ***Alzheimer's*** Pathway with Tau and BACE1 Inhibition

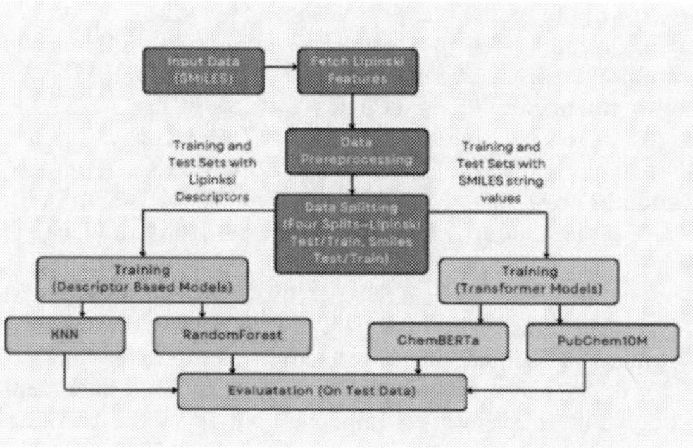

Figure 3. Research Methodology Flowchart

we present the results of our model evaluations, highlighting the performance of each approach. The implications of these results are discussed in the discussion and evaluation portions of this paper, where we also evaluate their significance in the context of Alzheimer's ***drug discovery***. The conclusion concludes the paper, offering insights into future work and potential improvements to the models.

II. Related work | Methods

The fundamental idea this research draws upon is that BACE1 is an effective inhibition target for Alzheimer's reduction. This idea was drawn upon by a previous paper. Gosh et al. proved BACE1 as a potential inhibition target by expressing its numerous advantages: BACE1 is a key target for ***Alzheimer's*** disease (AD) treatment due to its early role in amyloid-β (Aβ) production; the gene deletion of BACE1 produces only mild phenotypes, suggesting that inhibiting this enzyme might not have severe side effects; BACE1 is an aspartic protease, so the mechanism and inhibition of BACE1 are well-documented and researched, etc [14].

This research also fine-tunes and evaluates transformers that are trained upon molecular properties. Chithrananda et al. built such a model, ChemBERTa [15]. This model takes the well-known RoBERTa transformer [16] and fine-tunes it such that it can predict certain molecular properties. This research takes this model one step further and fine-tunes ChemBERTa to predict whether drugs can effectively inhibit BACE1. The RoBERTa transformer is indeed based on another transformer, BERT [17], which is based on the transformer architecture.

Similar research has been conducted; for example, Baressi et al., conducted research attempting to create models to predict the efficacy of medication on COVID-19 [18]. This research takes that idea further through the evaluation of drugs' efficacy on ***Alzheimer's*** while simultaneously comparing traditional models with newer transformer-based models. This therefore allowed for the evaluation of the difference in efficacy of the two types of models, opening the window for generalizations in this sphere of research.

A. Dataset

To conduct this research, data was collected from the CHEMBL4822 database (Figure 3—input data) that contained different drugs' Simplified Molecular Input Line Entry System (SMILES) notations - simple text-based representations of the drugs molecular structure [19] - paired with their half maximal inhibitory concentration (IC50) scores - a value indicating the dosage needed for a drug to effectively inhibit a certain protein, in this case BACE1. The dataset contains 10619 different drugs (including duplicates) and 46 additional descriptors with the focus being their IC50 scores.

B. Prepossessing

In the preprocessing stage, we first acquired data from the CHEML dataset and filtered it to retain entries where IC50 was specified as the standard type. Null values were addressed with mean imputation, replacing any null values with the dataset's average values, as cited in [20]. We also eliminated duplicate entries, resulting in a refined dataset comprising 7,353 distinct drugs. The data was then narrowed down to the 'canonical_smiles' and 'standard_value' columns. Here, 'canonical_smiles' represents the SMILES notation, and 'standard_value' corresponds to the IC50 value for each drug. To facilitate easier calculations and comparisons, IC50 scores were transformed into pIC50 [18], by taking the negative logarithm of IC50 in molar form. At this stage, our dataset contained two columns: SMILES and pIC50.

Subsequently, we extracted 210 Lipinski features for each SMILES notation entry, thereby expanding the dataset to include 211 columns while maintaining the 7,353 rows. This dataset was then divided into training and test subsets using an 8:2 ratio. For traditional ***machine learning*** models, we utilized the dataset with Lipinski features, whereas for transformer-based models, we retained only the SMILES and pIC50 values. This resulted in four distinct data files comprising training and test sets with Lipinski descriptors as well as training and test sets with SMILES notation. The training sets were employed to develop models using a 5-fold cross-validation approach, while the test sets were reserved to evaluate the models' predictive performance. This careful splitting was important to evaluate how good each model is thoroughly at making predictions.

C. Why Feature Extraction?

This research used two methods to analyze molecular structures: traditional models (KNN and RandomForest) and ***transformer models*** (ChemBERTa and PubChem10M). Traditional models cannot directly process SMILES notation, which represents molecules, ***transformer models*** were also used as they can handle SMILES values directly and may provide better results. For the traditional method, numerical values that describe the drug's properties were needed. 210 descriptors were extracted (such as HeavyAtomMolWt, ExactMolWt, NumValenceElectrons, NumRadicalElectrons, MolWt, etc.), referred to as Lipinski Descriptors [21], from the SMILES strings using the RDKit tool (see Figure 3 to see the data was processed) [22]. This allowed the traditional models to effectively analyze the molecules using these numerical values.

D. Training

After the dataset was both preprocessed and split, AI models were developed to predict drug effectiveness using PIC50 values: two descriptor-based models, KNN [23] and Random Forest [24], and two ***transformer models***, ChemBERTa and PubChem10M [25]).

We selected K-Nearest Neighbors (KNN), Random Forest, ChemBERTa, and PubChem10M models to leverage diverse analytical strengths. KNN and Random Forest are reliable, traditional models ideal for structured data and feature interpretability, providing a solid baseline with molecular descriptors. ChemBERTa and PubChem10M, as transformer-based models, excel in processing sequence data like SMILES strings, capturing complex molecular interactions more holistically. This combination of models allows us to comprehensively evaluate drug efficacy in inhibiting ***Beta-Secretase 1*** (BACE1), balancing robustness with innovative pattern recognition capabilities.

By exploring various hyperparameter settings, each model was trained to find the most optimal configurations. The Mean Squared Error (MSE) [26] and R-squared (R^2) [27] metrics were used to evaluate their performance. The model with the best results, determined by these metrics, was further tested on unseen data to ascertain its R^2 value. This process aimed to find the most suitable model for identifying drugs that might combat Alzheimer's disease, contributing to the discovery of potential treatments and benchmarking different modeling methods. The R^2 metric is a statistical measure often used to assess the accuracy of a regression task. Baressi et al. uses this metric when evaluating the accuracy of AI models to predict pIC50 values of COVID-19 medication [18].

III. Results

When evaluated, the four ***machine learning*** models—K-Nearest Neighbors (KNN), Random Forest, ChemBERTa,

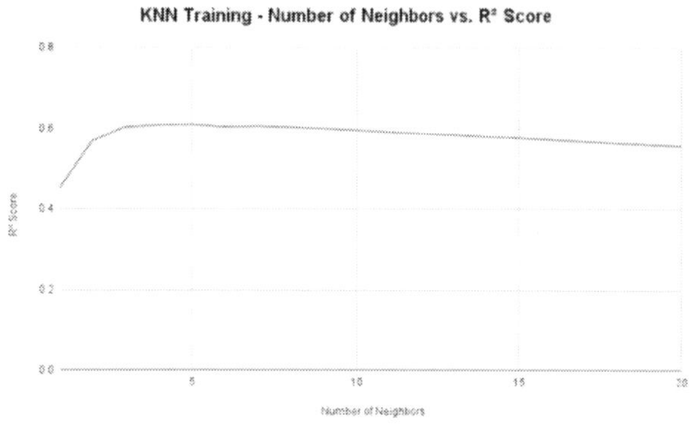

Figure 4. KNN Training Graph - Number of Neighbors vs. R^2 Score

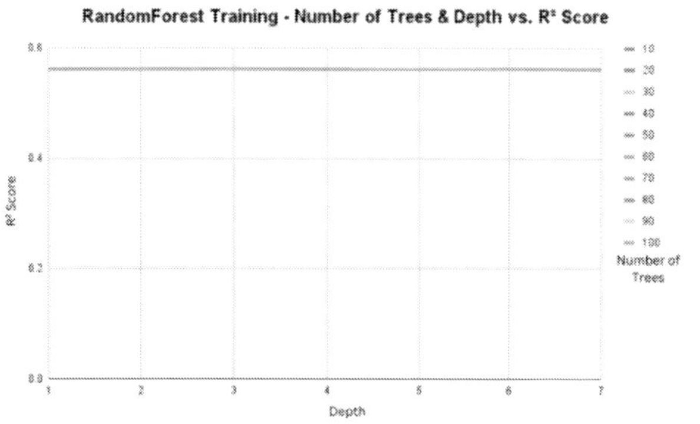

Figure 5. RandomForest Training Graph - Number of Trees and Depth vs. R^2 Score

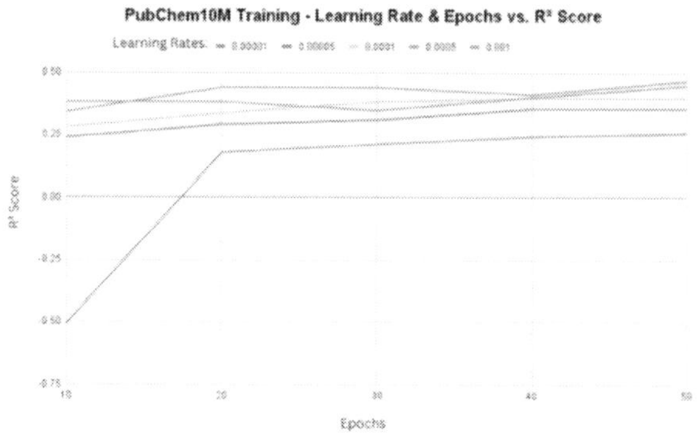

Figure 6. PubChem10M Training Graph - Learning Rate and Epochs vs. R^2 Score

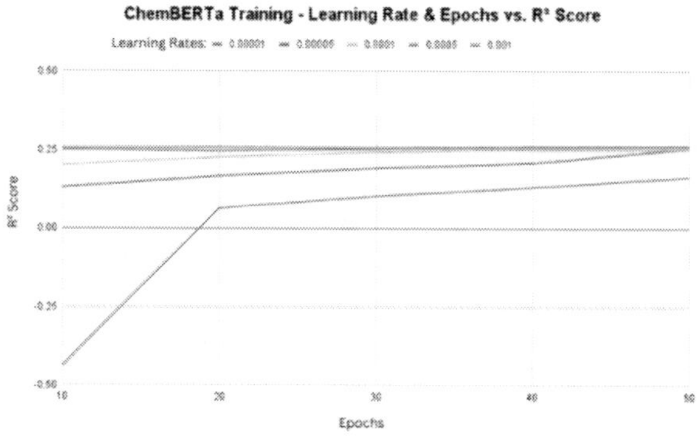

Figure 7. ChemBERTa Training Graph - Learning Rate and Epochs vs. R^2 Score

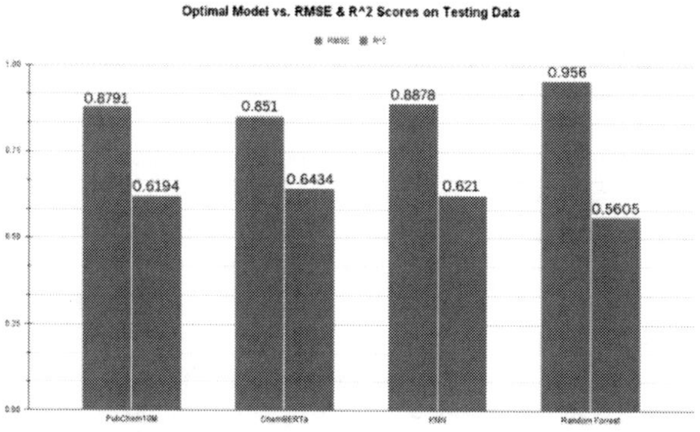

Figure 8. Optimal Model vs. RMSE and R^2 Scores on Testing Data

and PubChem10M—demonstrated notable differences in their ability to predict certain drugs' inhibition of BACE1. On training data, the KNN model (K=5) was the most effective, boasting the highest R^2 score of 0.6092 (see Figure 4). It was closely followed by the Random Forest model, with an R^2 of 0.5651 (maximum depth of 7 and 60 trees) (see Figure 5). In third place was the PubChem10M model (trained for 50 epochs with a learning rate of 0.001), achieving an R^2 score of 0.4672 (see Figure 6). Finally, in last place for training data, was the ChemBERTa model, with an R^2 score of 0.2641 (see Figure 7 and Table 1). However, on testing data, the models' results demonstrated a significant shift. The ChemBERTa model led the pack with an R^2 score of 0.6433, indicating strong generalization to unseen data. This was closely followed by the KNN model with an R^2 score of 0.621—a continuation of its robust performance in training. The PubChem10M model also showed substantial improvement, achieving an R^2 score of 0.6194. Lastly, the Random Forest model displayed great consistency, scoring an R^2 of 0.5605 (see Table 1).

The observed trends underscore the architectural advantages of transformers, such as ChemBERTa and PubChem10M. These models excel due to the attention mechanism, which enables them to capture and generalize complex data patterns inherent

TABLE I
RESULT SUMMARY

Model	Training R^2	Testing R^2
KNN	0.609	0.621
Random Forest	0.565	0.560
ChemBERTa	0.264	0.643
PubChem10M	0.467	0.619

in molecular SMILES notation—even when initial training R^2 scores are low. We conducted a series of experiments to better understand these findings, analyzing cross-validation R^2 scores and performance metrics across multiple data subsets. Consistently, the results demonstrated that ***transformer models*** achieve higher test R^2 scores, confirming their superior ability to generalize under varied conditions compared to traditional ***machine learning*** models.

The ChemBERTa model was able to perform well on testing data among metrics boasting the lowest RMSE score of about 0.851 (see Figure 8).

IV. DISCUSSION | EVALUATION

This study contained the evaluation of the KNN, Random-Forest, ChemBERTa and PubChem10M models in predicting drugs' ability to inhibit the Alzheimer's pathway using the R^2 score as a metric for the accuracy and effectiveness of each model.

The KNN model performed the best on training data with an R^2 score of 0.6092 with a parameter setting of K=5 neighbors. This indicates the model was proficient in fitting the training data when using K=5 neighbors. In the testing phase, this R^2 score stayed relatively consistent at 0.621.

The Random Forest model had a moderate R^2 score of 0.5651 during training with a maximum depth of 7 and 60 estimators. Like the KNN model, it had a relatively consistent score during testing of 0.5605. This indicates that both the KNN and RandomForest models were relatively consistent models. For the RandomForest model, this consistency might be due to the ensemble nature of Random Forest, which averages multiple decision trees to achieve a robust prediction, making it less likely to be affected by variance in the data [28].

ChemBERTa, using 50 epochs and a learning rate of 0.001, achieved a relatively low R^2 score of 0.2641 during training. That said, the model significantly improved when evaluated on testing data, achieving the highest R^2 score of 0.6433 among all the models. The massive increase in R^2 score suggests that ChemBERTa is particularly good at learning patterns during training and by extension effectively generalizes new data. It shows a large potential for the model to capture the underlying data distribution regardless of initially low training performance.

The PubChem10M model, also trained for 50 epochs with a learning rate of 0.001, had a moderate R^2 score of 0.4672 during training. Like ChemBERTa, the PubChem10M model showed a significant improvement in the testing phase with an R^2 score of 0.6194. This improvement indicates that the PubChem10M model, although not as strong as ChemBERTa, has robust generalization properties. It has comparable testing performance to that of KNN, despite a low initial training score.

Each model showcased unique characteristics across the datasets. The KNN model excelled during training but displayed inconsistencies in testing. The Random Forest model maintained consistent performance across both datasets, but it did not achieve the high levels of accuracy seen in other models. ChemBERTa showed the most notable improvement across the phases. The incorporation of both traditional and newer transformer-based models allows for this research to effectively create generalizations that are lacking in preexisting research.

The ChemBERTa and PubChem10M models, on the other hand, may have built robust patterns during the training phase that were solid and applicable to the testing data. This could be extremely powerful if refined even further.

These results offer valuable insights into the strengths and weaknesses of each model, guiding future research and practical applications where different models may be more suited depending on the context and the nature of the data.

Since ensemble models are hypothesized to have provided consistency and transformers for greater generalization, one could experiment with combining the ensemble nature of the RandomForest model with a transformer to obtain greater results. One could also expand the number of epochs and learning rate values tested to see if there are model configurations that generate better results.

It is important to note that these models do not account for bioavailability, pharmacokinetics, or potential interactions with other drugs, which are critical factors for clinical outcomes. These aspects are essential for understanding how a drug behaves in the body and how effective it will be in real-world scenarios. Consequently, while the models offer insights into drug potential, a more comprehensive approach that includes these factors is necessary for enhancing clinical relevance.

V. CONCLUSION AND FUTURE WORK

This study documented the evaluation of four distinct AI-based models—K-Nearest Neighbors (KNN), Random Forest, ChemBERTa, and PubChem10M—in their ability to successfully estimate how effectively a certain drug could disrupt the Alzhimer's pathway. In this evaluation, results within the testing and training phase were quite varied when evaluating ChemBERTa and PubChem10M models: both models showed low, unfavorable R^2 scores, yet, when these models got to the testing phase, their scores increased by a large margin boasting R^2 scores of 0.6433 (the highest testing score among all the models) and 0.6194 respectively. This indicated unique generalization prowess among the transformer-based models. The descriptor-based models—KNN and RandomForest—on the other hand, were pretty stable; the KNN model had an R^2 score of 0.6092 during training and an R^2 score of 0.621 within testing—indicating powerful consistency. This same trend applied to the RandomForrest model which had a lower R^2 score of 0.5651 but it stayed pretty consistent reaching 0.5605 during testing. These results imply that the transformer-based models had powerful generalization capabilities whereas the descriptor-based models boasted consistency. This highlights the age-old accuracy vs. precision problem which is present in our study today. The main limitation of this study is that it only looks at IC50 values to assess drug potency. While IC50 is important, it does not fully reflect how a drug will work in real life because factors like how the drug is absorbed, distributed, metabolized, and excreted (ADME), and its toxicity also play a role. Our models do not take into account how drugs interact with the body, which might lead to differences between predicted results and actual effects. This study should be seen as a starting point, and future work should include these other factors to make the models more useful for real-world drug development.

REFERENCES

[1] K. Blennow, M. J. de Leon, and H. Zetterberg, "Alzheimer's disease", *The Lancet*, vol. 368, no. 9533, pp. 387–403, 2006.

[2] W. Thies and L. Bleiler, "Alzheimer's disease facts and figures alzheimer's association", *Alzheimers Dement*, vol. 8, no. 2, pp. 131–168, 2012.

[3] J. S. Birks and R. J. Harvey, "Donepezil for dementia due to alzheimer's disease", *Cochrane Database of systematic reviews*, no. 6, 2018.

[4] J. S. Birks and J. G. Evans, "Rivastigmine for alzheimer's disease", *Cochrane Database of systematic reviews*, no. 4, 2015.

[5] B. Jarvis and D. P. Figgitt, "Memantine", *Drugs & aging*, vol. 20, pp. 465–476, 2003.

[6] G. E. Vitek, B. Decourt, and M. N. Sabbagh, "Lecanemab (ban2401): An anti–beta-amyloid monoclonal antibody for the treatment of alzheimer disease", *Expert opinion on investigational drugs*, vol. 32, no. 2, pp. 89–94, 2023.

[7] R. H. Takahashi, T. Nagao, and G. K. Gouras, "Plaque formation and the intraneuronal accumulation of β-amyloid in alzheimer's disease", *Pathology international*, vol. 67, no. 4, pp. 185–193, 2017.

[8] R. J. O'brien and P. C. Wong, "Amyloid precursor protein processing and alzheimer's disease", *Annual review of neuroscience*, vol. 34, no. 1, pp. 185–204, 2011.

[9] E.-M. Mandelkow and E. Mandelkow, "Tau in alzheimer's disease", *Trends in cell biology*, vol. 8, no. 11, pp. 425–427, 1998.

[10] G. S. Bloom, "Amyloid-β and tau: The trigger and bullet in alzheimer disease pathogenesis", *JAMA neurology*, vol. 71, no. 4, pp. 505–508, 2014.

[15] S. Chithrananda, G. Grand, and B. Ramsundar, "Chemberta: Large-scale self-supervised pretraining for molecular property prediction", *arXiv preprint arXiv:2010.09885*, 2020.

[16] Y. Liu *et al.*, "Roberta: A robustly optimized bert pretraining approach", *arXiv preprint arXiv:1907.11692*, 2019.

[11] K. Basu, R. Sinha, A. Ong, and T. Basu, "Artificial intelligence: How is it changing medical sciences and its future?", *Indian journal of dermatology*, vol. 65, no. 5, pp. 365–370, 2020.

[12] A. Vaswani, "Attention is all you need", *arXiv preprint arXiv:1706.03762*, 2017.

[13] J. Born and M. Manica, "Regression transformer enables concurrent sequence regression and generation for molecular language modelling", *Nature Machine Intelligence*, vol. 5, no. 4, pp. 432–444, 2023.

[14] A. K. Ghosh, S. Gemma, and J. Tang, "β-secretase as a therapeutic target for alzheimer's disease", *Neurotherapeutics*, vol. 5, pp. 399–408, 2008.

[17] J. Devlin, "Bert: Pre-training of deep bidirectional transformers for language understanding", *arXiv preprint arXiv:1810.04805*, 2018.

[18] S. Baressi Šegota, I. Lorencin, Z. Kovač, and Z. Car, "On approximating the pic 50 value of covid-19 medicines in silico with artificial neural networks", *Biomedicines*, vol. 11, no. 2, p. 284, 2023.

[19] D. Weininger, "Smiles, a chemical language and information system. 1. introduction to methodology and encoding rules", *Journal of chemical information and computer sciences*, vol. 28, no. 1, pp. 31–36, 1988.

[20] A. R. T. Donders, G. J. Van Der Heijden, T. Stijnen, and K. G. Moons, "A gentle introduction to imputation of missing values", *Journal of clinical epidemiology*, vol. 59, no. 10, pp. 1087–1091, 2006.

[21] H. Mishra, N. Singh, T. Lahiri, and K. Misra, "A comparative study on the molecular descriptors for predicting drug-likeness of small molecules", *Bioinformation*, vol. 3, no. 9, p. 384, 2009.

[22] G. Landrum, "Rdkit documentation", *Release*, vol. 1, no. 1-79, p. 4, 2013.

[23] M. Steinbach and P.-N. Tan, "Knn: K-nearest neighbors", in *The top ten algorithms in data mining*, Chapman and Hall/CRC, 2009, pp. 165–176.

[24] S. J. Rigatti, "Random forest", *Journal of Insurance Medicine*, vol. 47, no. 1, pp. 31–39, 2017.

[25] Y. Omote, K. Matsushita, T. Iwakura, A. Tamura, and T. Ninomiya, "Transformer-based approach for predicting chemical compound structures", in *Proceedings of the 1st Conference of the Asia-Pacific Chapter of the Association for Computational Linguistics and the 10th International Joint Conference on Natural Language Processing*, 2020, pp. 154–162.

[26] M. S. Error, "Mean squared error", *MA: Springer US*, pp. 653–653, 2010.

[27] A. Gelman, B. Goodrich, J. Gabry, and A. Vehtari, "R-squared for bayesian regression models", *The American Statistician*, 2019.

[28] M. A. Ganaie, M. Hu, A. K. Malik, M. Tanveer, and P. N. Suganthan, "Ensemble deep learning: A review", *Engineering Applications of Artificial Intelligence*, vol. 115, p. 105 151, 2022.

A Hybrid Machine Learning Approach for Enhanced Skin Cancer Diagnosis Using Convolutional Neural Networks, Support Vector Machines, and Gradient Boosting

Fazila Faizal Patel
School of Engineering, Technology and Design
Canterbury Christ Church University
Canterbury, United Kingdom
E-mail: f.patel149@canterbury.ac.uk

Amina Souag
School of Engineering, Technology and Design
Canterbury Christ Church University
Canterbury, United Kingdom
E-mail: amina.souag@canterbury.ac.uk

Adedayo Olowolayemo
School of Engineering, Technology and Design
Canterbury Christ Church University
Canterbury, United Kingdom
E-mail: a.olowolayemo502@canterbury.ac.uk

***Abstract*— This study investigates the effectiveness of a hybrid machine learning model for skin cancer diagnosis, integrating Convolutional Neural Networks, Support Vector Machines, and Gradient Boosting algorithms. By combining the strengths of each technique, the model seeks to improve diagnostic accuracy and reliability in clinical settings, addressing the challenges posed by traditional diagnostic methods. Utilizing the "Skin Cancer: Malignant vs. Benign" dataset, the hybrid model achieved an accuracy of 84%, with precision, recall, F1 score, and specificity recorded at 85%, 84%, 84%, and 83%, respectively. These results underscore the model's potential to surpass single-algorithm approaches in detecting skin cancer, making it a promising tool for early diagnosis and better-informed clinical decision-making. The findings highlight the broader impact of advanced machine learning techniques in healthcare, particularly in oncology, by demonstrating how the integration of multiple algorithms can provide more accurate, scalable, and reliable diagnostic solutions. This research opens avenues for further exploration of hybrid models as a means to advance AI-driven diagnostic technologies in medical fields, with potential applications across various types of cancer detection. The source code for this study is available through a public GitHub repository, fostering transparency and further innovation in the field.**

Keywords- Hybrid Machine Learning; Skin Cancer; Convolutional Neural Networks; Support Vector Machines; and Gradient Boosting.

I. INTRODUCTION

Cancer continues to be a pressing global health concern, accounting for a significant share of mortality rates around the world [1][2]. Early detection and accurate diagnosis are vital components in enhancing patient outcomes [3][4], particularly in the case of skin cancer, where timely interventions can lead to markedly better survival chances [5]. While established diagnostic methods, such as visual inspection, biopsy, and histopathology have their merits, they are often susceptible to human error and subjectivity in interpretation [6]. These limitations highlight the urgent need for more reliable and automated diagnostic tools that can support healthcare professionals in making consistent and accurate diagnoses.

In the realm of healthcare, Machine Learning (ML) has emerged as a transformative force, providing innovative ways to enhance diagnostic accuracy by efficiently analyzing large and complex datasets. Various ML models, including Convolutional Neural Networks (CNNs) and Support Vector Machines (SVMs), have been successfully applied in cancer diagnosis [7][8]. However, each model comes with its own set of challenges. For instance, while CNNs excel at extracting meaningful features from image data, their effectiveness can diminish when training datasets are insufficient. On the other hand, SVMs are proficient at managing high-dimensional data but may struggle with scalability when confronted with larger datasets [9]. These nuances highlight the necessity of refining diagnostic methodologies to fully harness the potential of machine learning.

Hybrid machine learning models have gained prominence as a solution to the inherent limitations of individual techniques, leveraging the complementary strengths of multiple algorithms to enhance performance across a range of applications [10]. While hybridization itself is well-established, the contribution of this research lies in the sophisticated fusion of CNNs, SVMs, and Gradient Boosting (GB), each selected for their distinct advantages in the context of skin cancer diagnosis. CNNs are employed for their unparalleled ability to autonomously extract hierarchical features from complex image data, effectively capturing spatial patterns that are crucial for medical image analysis. SVMs, known for their robustness in high-dimensional spaces, are used to classify these features accurately, especially in cases with intricate decision boundaries. GB, recognized for its ensemble learning capability, is integrated to refine predictive accuracy, enhance model generalization, and mitigate the risks of overfitting.

The innovation in this approach extends beyond the selection of individual algorithms; it lies in their seamless integration through advanced fusion strategies, such as weighted averaging and voting. These techniques harmonize the outputs of the models, optimizing the decision-making process and enhancing the overall reliability of predictions. This methodological synergy not only strengthens the model's resilience to variations in the data but also ensures a robust and scalable solution for skin cancer diagnosis.

The effectiveness of the hybrid framework is demonstrated through its superior performance across key metrics, including accuracy, precision, recall, F1 score, and specificity, compared to models based on a single algorithm. This research provides a detailed exploration of the methodology, experimental design, and results, underscoring the potential of hybrid models to significantly advance diagnostic capabilities. By strategically combining CNNs, SVMs, and GB, this approach offers a novel solution that not only leverages the strengths of each algorithm but also mitigates their individual weaknesses, establishing a compelling contribution to the field of medical diagnostic systems.

The structure of the rest of this paper is organized as follows: Section II explores related works, emphasizing the various machine learning approaches employed in cancer diagnosis, hybrid algorithms used in cancer research, and the data types and sources referenced in prior studies. Section III outlines our research approach in detail, including descriptions of data collection and preprocessing, model development, and methodological flowcharts that illustrate the study's workflow. Section IV presents our findings, supported by a detailed analysis and contextual discussion to interpret their significance. Finally, Section V summarizes the key insights of the study, highlights its limitations, and offers directions for future research.

II. LITERATURE REVIEW

Prompt and accurate identification is crucial for effective treatment and improved patient results. Conventional methods for diagnosing cancer, including imaging, histology, and genetic testing, are restricted in their ability to accurately detect, identify, and understand the progression of the disease. Studies have shown that advancements in machine learning have significantly impacted various sectors, including healthcare and these developments have facilitated the design and implementation of automated diagnostic tools, which have demonstrated improved accuracy in specific applications, as evidenced by numerous studies and clinical trials [11].

A. *Machine Learning in Cancer Diagnosis*

There are four main types of machine learning: supervised, unsupervised, semi-supervised, and reinforcement learning. Supervised learning involves training models on labeled data for tasks like classification and regression, while unsupervised learning identifies patterns and structures within unlabeled data, often through clustering techniques. Semi-supervised learning combines a small amount of labeled data with a larger pool of unlabeled data to improve model performance, and reinforcement learning teaches agents to make sequential decisions by rewarding desired behaviours and penalizing undesired ones.

To support these types of learning, preprocessing techniques like feature selection and normalization are essential, as they improve model effectiveness by selecting the most relevant features and ensuring that data is appropriately scaled. These steps contribute to higher model accuracy and efficiency, particularly when dealing with diverse datasets across machine-learning approaches. In the field of cancer diagnosis, where data is often complex and high-dimensional, such preprocessing methods play a crucial role in optimizing model performance. Hybrid models, in particular, benefit from combining classification and clustering techniques to handle different aspects of the data [12]. For instance, cancer diagnosis hybrid models incorporate both classification and clustering methods and often integrate feature selection and dimensionality reduction techniques to further enhance diagnostic accuracy and efficiency.

Cluster-based classification involves using clustering algorithms to organize data before training a classifier, which helps to reduce variation within clusters and enhance overall performance. Combining predictions from various models like SVM, Random Forest, and Neural Networks in ensemble approaches enhances both accuracy and robustness. In addition, deep learning with feature engineering involves utilising deep learning models to extract features directly from raw data like images or genetic sequences and inputting them into standard classifiers for improved prediction [13].

As Artificial Intelligence (AI) and ML quickly grow in importance, they're becoming essential tools in healthcare, especially for diagnosing and treating diseases. Cancer diagnosis is a prime example where these technologies can make a real difference [14]. Growing evidence shows that detecting cancer at an early stage leads to better treatment options and significantly improved patient outcomes. Consequently, researchers have leveraged hybrid machine learning models, combining multiple algorithms to capitalize on their individual strengths and compensate for their limitations, thus enhancing diagnostic accuracy and robustness in cancer detection [15][16]. In cancer detection, for instance, ensemble techniques like bagging, boosting, or stacking are commonly used to blend predictions from multiple models, thereby enhancing accuracy and reliability. Deep learning models, such as CNNs, are often paired with traditional algorithms like SVMs, creating a robust approach especially valuable for complex data, such as medical imaging [17].

An additional strength of these hybrid models lies in their ability to integrate diverse data types which may include genomics, imaging, and clinical records into a unified diagnostic tool. By drawing on multiple sources, hybrid models can better capture the complex biological patterns associated with cancer. For example, logistic regression and elastic net techniques are sometimes applied to classify genetic variations with higher risks, while specialized tools like LungCLiP use ensemble classifiers to detect lung cancer in plasma samples [18]. Altogether, these integrative machine learning approaches are advancing diagnostic accuracy, supporting earlier detection, and enabling more personalized treatment strategies in cancer care [19].

B. Hybrid Algorithms in Cancer Studies

Hybrid algorithms in cancer research employ various techniques to enhance the precision and effectiveness of cancer diagnosis, treatment planning, and prognosis prediction. Selecting the right hybrid approach depends on multiple factors, including the type of cancer, the nature of available data, specific diagnostic objectives, and desired outcomes [20][21]. CNNs with SVMs for example takes advantage of both approaches' strengths: CNNs' excellent feature extraction capabilities and SVMs' ability to classify small numbers of data [22]. Saleh et al. [22] used a hybrid CNN-SVM method for classifying lung CT images into four categories: adenocarcinoma, large cell carcinoma, normal, and squamous cell carcinoma. The method was tested on the Chest CT-Scan images dataset, achieving a high classification accuracy rate of 97.91%, which outperformed other recent deep learning-based works.

The method used by Saleh et al. [22] also demonstrated promising performance in terms of sensitivity, specificity, precision, and AUC. The results suggest that the hybrid CNN-SVM method has the potential to assist in the early detection of lung cancer, and future work can focus on testing the method with different datasets and image types [22]. Combining CNN and Recurrent Neural Networks (RNNSs) for cancer diagnosis as demonstrated in [23] takes advantage of the capabilities of each architecture and in [8] where CNNs extract spatial features, whereas RNNs recognize temporal or sequential patterns. This hybrid technique is particularly useful for evaluating complicated medical data, such as time-series medical pictures or sequential patient records.

These studies highlight some of the advancements in cancer diagnosis through hybrid machine learning models. By combining the strengths of different algorithms, such as CNNs with SVMs or RNNs, hybrid approaches have demonstrated improved performance in various cancer detection tasks, achieving higher accuracy and reliability compared to traditional single-algorithm methods. Table I shows reviews of related works that use Hybrid Machine Learning for Cancer Diagnosis.

TABLE I. REVIEW OF OTHER RELATED STUDIES

Article Ref.	Data Source	Records	Train/Test Split	Algorithm Type	Model Accuracy
[22]	Chest CT-Scan Images	5103	80:20	CNN, SVM	97.91%
[29]	Herlev public	917	80:20	CNN, SVM	99.30%
[31]	Breast Cancer Network Wisconsin	-	70-30	K-means, SVM	97.34%
[30]	Mammographic Image Analysis Society	-	70-30%	CNN, GRU	95.50%
[32]	PCAM Kaggle	277524	80:20	CNN, GRU	86.21%
[23]	-Lung Image Database Consortium and Image Database Resource Initiative (LIDC-IDRI)	888	80:20	CNN, RNN	95.00%

C. Datasets and Data Sources

Datasets are the foundation of machine learning, essential for training, validating, and testing models to ensure robustness, generalizability, and the ability to tackle real-world challenges effectively. A well-curated, diverse, and high-quality dataset is crucial to the success of any machine learning effort, as it enables models to perform accurately while minimizing biases. For example, Yogendra Singh Solanki et al. [24] developed an ML-based classifier system for breast cancer prognosis using a dataset from the University of California, Irvine (UCI) repository to distinguish between malignant and benign breast cancer cells. In building such models, data imbalance often poses a significant challenge, as it can skew predictions toward the more prevalent class. Yogendra Singh Solanki et al. addressed this by using techniques like re-sampling and the Synthetic Minority Over-sampling Technique (SMOTE), a method for handling class imbalance by generating new realistic samples for the minority class, which helps models learn more evenly and reduces bias toward the majority class.

In their study, Wang et al. [25] obtained cancer data from three distinct data sources to analyze cancer incidences, which include, The Cancer Genome Atlas (TCGA), Surveillance, Epidemiology, and End Results (SEER) 18, and North American Association of Central Cancer Registries. The TCGA shared information on individuals with 33 different forms of cancer, using specific TCGA case IDs to prevent any repeat cases among the different types of cancer. SEER data includes individuals who were diagnosed with primary cancer between 2010 and 2013, characterized by the third edition of the International Classification of Diseases for Oncology (ICD-O-3) using primary site and histology/behavior criteria. The NAACCR database included cancer records from every state in the US as well as the District of Columbia, encompassing nearly the entire population of the country from 2009 to 2013.

In order to analyze the distribution of races, only cancer cases in the US with race information were used, taking into account the SEER program's overrepresentation of minority populations in the US. These expansive data collections cover 33 different types of cancer and can be identified by distinct case IDs and ICD-O-3 categorization, offering extensive population representation and valuable insights into cancer case characteristics in the US [25].

III. METHODOLOGY

This paper focuses on developing a robust hybrid machine learning model for cancer diagnosis, encompassing several key phases: data collection and preprocessing, exploratory data analysis, feature engineering, model selection, and implementation following our methodology framework as in Figure 1. Utilizing the "Skin Cancer: Malignant vs. Benign" image dataset, exploratory data analysis provided insights into the dataset composition, guiding feature engineering, which tailored the data for use in CNNs, SVMs, and GB models. Model selection and implementation involved carefully combining these techniques using Python and Scikit-Learn, optimizing the model's performance through hyperparameter tuning and an 80/20 train-test split. The results demonstrated the model's effectiveness, offering promising advancements in accurate cancer diagnosis.

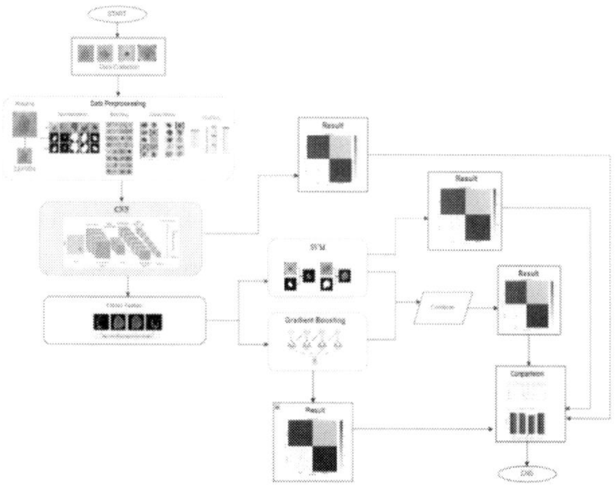

Figure 1. Research methodology flowchart.

A. *Data Collection and Preprocessing*

The "Skin Cancer: Malignant vs Benign" dataset [26], used in this study, consists of 3,600 images of skin moles, evenly divided between benign (1,800) and malignant (1,800) cases. Each image, sourced from the ISIC Archive, a well-regarded repository for dermatological images is provided at a consistent resolution of 224x244 pixels, ensuring uniform quality for analysis. This balanced dataset with samples shown in Figure 2 below serves as an essential resource for developing and validating machine learning models to improve the accuracy and reliability of skin cancer diagnosis.

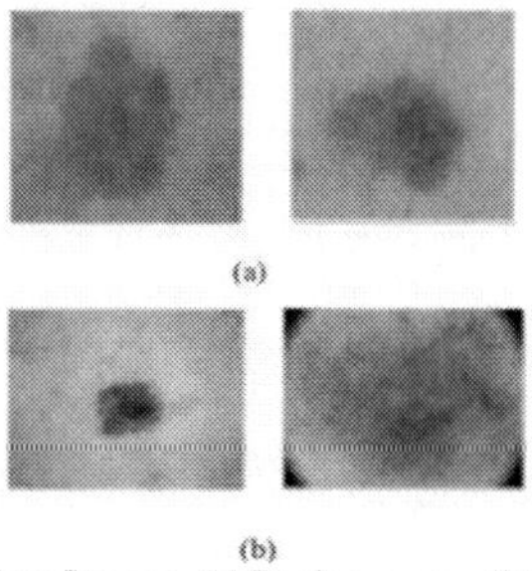

Figure 2. Skin cancer Images- (a) Benign cancer, (b) Malignant cancer

With a training set of 2,637 images (1,440 benign and 1,197 malignant) and a test set of 963 images (360 benign and 300 malignant), split into an 80/20 ratio for training and testing, this dataset provides a robust foundation for training and evaluating hybrid machine learning models for cancer diagnosis.

The boxplot in Figure 3 and Figure 4 visualizes the distribution of the "mean intensity" feature for benign and malignant tumor images based on the training and test data respectively, highlighting the differences in typical values (median, quartiles) and variability (spread) between the two classes. Outliers, represented as individual circles outside the whiskers, indicate data points that fall significantly outside the general distribution. This suggests that mean intensity could be a valuable feature for classifying tumors, as benign tumors generally show lower mean intensity compared to malignant ones. Additionally, this feature could be used directly in machine learning models or serve as a basis for creating more complex features. Handling outliers may be necessary to avoid skewing model training, and understanding the distribution of mean intensity can inform decisions on data preprocessing and feature engineering strategies.

The images were resized to 224x224 pixels to ensure uniformity and compatibility with CNN. The pixel values are normalized to a range between 0 and 1 to improve model convergence during training. The dataset is also shuffled to prevent overfitting, ensuring that the model learns from a balanced and randomized distribution of benign and malignant cases.

CNNs are employed for automatic feature extraction from the image data. In parallel, the Gabor filters and Gray-Level Co-occurrence Matrix (GLCM) are applied for texture-based feature extraction. The extracted features from both methods are then combined through feature fusion to enhance classification capabilities. Principal Component Analysis (PCA) was then used to reduce the dimensionality of the combined feature set. This step was to ensure that the most important features are retained while reducing computational complexity, leading to a more efficient and scalable model.

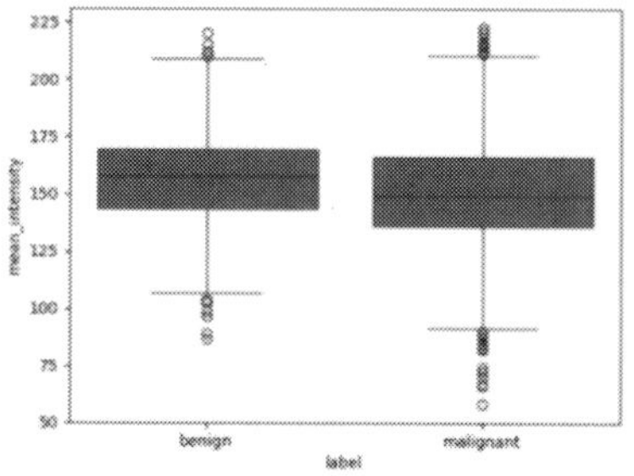

Figure 3. Boxplot of mean intensity by label (Training Data)

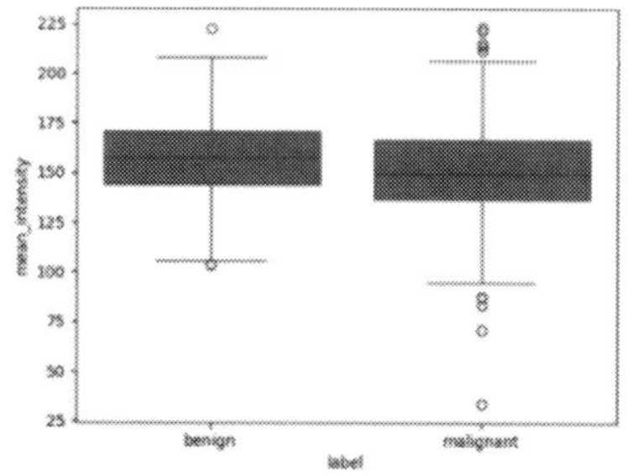

Figure 4. Boxplot of mean intensity by label (Test Data)

B. *Model Development*

This study develops a robust hybrid machine learning model for skin cancer diagnosis, combining the strengths of CNNs, SVMs, and GB. The model's core design is based on leveraging the complementary capabilities of these algorithms: CNNs for automatic feature extraction from skin cancer images, SVMs for efficient classification in high-dimensional feature spaces, and GB for ensemble learning that refines predictive performance by aggregating weak learners into a more robust model. Specifically, CNNs excel at identifying intricate spatial patterns within the images, while SVMs handle complex decision boundaries in the high-dimensional feature space, and GB enhances generalization by reducing overfitting.

The integration of these algorithms is facilitated by advanced fusion strategies, such as weighted averaging and voting, which harmonize the individual outputs to optimize decision-making. This methodological synergy ensures that the model's predictive reliability is enhanced, particularly in clinical settings where diagnostic accuracy is paramount. The integration process is sequential, with CNNs first extracting relevant features, followed by SVM classification and GB aggregation, to ensure a comprehensive approach to skin cancer diagnosis.

The model was implemented using Python 3.12.7 and the following libraries: Scikit-learn 1.4.2 for SVM and GB, Keras/TensorFlow for building and training the CNN, and Pandas, NumPy, Matplotlib, and Seaborn for data manipulation, feature engineering, and visualization. The "Skin Cancer: Malignant vs. Benign" dataset, sourced from the ISIC Archive, comprises 3,600 images, evenly distributed between benign and malignant cases. Images are resized to 224x224 pixels and normalized to improve model convergence during training. Data preprocessing includes shuffling to prevent overfitting and feature engineering through Gabor filters and Gray-Level Co-occurrence Matrix (GLCM), with dimensionality reduction applied using PCA to maintain computational efficiency while retaining essential features.

Hyperparameter tuning was performed on each model component: CNN layers, SVM parameters, and GB hyperparameters, using grid search and cross-validation to optimize performance. Evaluation metrics, including accuracy, precision, recall, F1 score, and specificity, were calculated and compared to single-algorithm models to demonstrate the efficacy of the hybrid approach. The confusion matrix was also generated to visualize the model's classification performance in terms of true positives, false positives, true negatives, and false negatives.

The dataset was partitioned into an 80% training set and a 20% testing set, ensuring that the model underwent thorough training while maintaining an unbiased and reliable evaluation process. This division facilitates both robust model learning and accurate performance assessment. The hybrid model's design effectively addresses several critical challenges commonly encountered in medical diagnostics, such as data sparsity, class imbalance, and computational scalability. By leveraging the strengths of multiple algorithms, the model not only mitigates the limitations of individual approaches but also achieves superior diagnostic performance, outperforming single-algorithm models in terms of accuracy and generalization. The source code for the implementation is publicly accessible via a GitHub repository [27], fostering transparency and providing an avenue for further research and development within the domain of skin cancer detection. This open-access model serves as a valuable resource for advancing the field and promoting collaborative exploration of hybrid machine learning techniques for medical image analysis.

IV. Results and Discussion

In this study, a thorough assessment and comparison of various machine learning models for cancer detection were carried out using SciKit-Learn. Table II shows the performance metrics Accuracy, Precision, Recall, F1-score, and Specificity for different models including the proposed hybrid model, CNN, SVM, and GB.

TABLE II. ML MODELS PERFORMANCES DATA

Metric	Hybrid Model	CNN	SVM	GB
Accuracy	84	82	74	83
Precision	85	84	74	82
Recall	84	82	74	85
F1	84	82	74	83
Specificity	83	73	78	80

Table III below shows confusion matrices for the Hybrid, CNN, GB, and SVM models illustrating the breakdown of True Positives (TP), False Positives (FP), True Negatives (TN), and False Negatives (FN) values essential for evaluating a model's classification performance by breaking down its accurate and inaccurate predictions, offering a comprehensive view of its classification abilities [28] while Figure 5 offers a comparison of these performance measures among the various models.

The findings of this study underscore the significance of the hybrid machine learning model in the field of skin cancer diagnostics. The model demonstrated high accuracy and robust performance metrics, suggesting its potential utility as a reliable tool for the early detection of skin cancer.

TABLE III. CONFUSION MATRIX OF PROPOSED HYBRID MODEL, CNN, GB, AND SVM MODELS.

Proposed Hybrid Model

Actual Value	Predicted Value: Positive	Predicted Value: Negative
TRUE	299	42
FALSE	61	258

CNN

Actual Value	Predicted Value: Positive	Predicted Value: Negative
TRUE	262	22
FALSE	98	278

SVM

Actual Value	Predicted Value: Positive	Predicted Value: Negative
TRUE	279	92
FALSE	81	208

Gboost

Actual Value	Predicted Value: Positive	Predicted Value: Negative
TRUE	279	50
FALSE	61	250

By properly evaluating skin cancer photos and utilizing the characteristics of many algorithms, the model has proved its capacity to reliably discern between benign and malignant skin moles. This is a step forward in the development of more precise diagnostic tools, perhaps leading to earlier identification and improved treatment outcomes for patients.

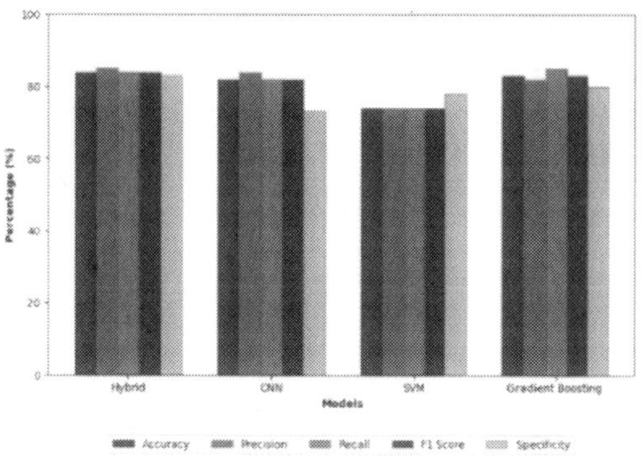

Figure 5. Visualization of the model's performances

The findings of this study underscore the significance of the hybrid machine learning model in the field of skin cancer diagnostics. The model demonstrated high accuracy and robust performance metrics, suggesting its potential utility as a reliable tool for the early detection of skin cancer. By properly evaluating skin cancer photos and utilizing the characteristics of many algorithms, the model has proved its capacity to reliably discern between benign and malignant skin moles. This is a step forward in the development of more precise diagnostic tools, perhaps leading to earlier identification and improved treatment outcomes for patients.

The model's strong performance in skin cancer diagnosis shows that it might be useful in clinical settings, allowing Medical practitioners make better informed decisions and eliminate diagnostic mistakes. Overall, the model's contribution to improving skin cancer detection is important for successful treatment and increasing patient survival rates.

V. Conclusion

This research focused on evaluating the effectiveness of a hybrid machine learning model for skin cancer diagnosis compared to traditional models like CNN, SVM, and GB. The hybrid model demonstrated superior performance with an accuracy of 84%, precision of 85%, recall of 84%, F1 score of 84%, and specificity of 83%. These metrics indicate that the hybrid model not only performed better than the SVM model, which had the lowest accuracy at 74%, but also outperformed the CNN and Gradient Boosting models in most aspects, particularly in specificity. The CNN model, while achieving high precision (84%) and accuracy (82%), lagged in specificity (73%), indicating a higher rate of false positives. In contrast, the Gradient Boosting model, with metrics closely matching the hybrid model, also showed strong performance but was slightly less effective overall.

This summary highlights the hybrid model's potential as a more reliable and accurate tool for skin cancer diagnosis. This paper is primarily focused on comparing single-algorithm models to hybrid Machine Learning models, however for future research directions, the proposed hybrid models can be compared with other potential hybrid models.

References

[1] H. Sung et al., "Global cancer statistics 2020: GLOBOCAN Estimates of Incidence and Mortality Worldwide for 36 Cancers in 185 Countries," CA Cancer J Clin, vol. 71, pp. 209–249, May 2021, doi: 10.3322/caac.21660.

[2] M. Dildar et al., "Skin cancer detection: A review using deep learning techniques," Int. J. Environ. Res. Public Health, vol. 18, pp. 5479, May 2021, doi: 10.3390/ijerph18105479.

[3] Y. Xie et al., "Early lung cancer diagnostic biomarker discovery by machine learning methods," Translational oncology., vol. 14, pp100907, Jan. 2021, doi: 10.1016/j.tranon.2020.100907.

[4] X. Guan et al., "Construction of the XGBoost model for early lung cancer prediction based on metabolic indices," BMC medical informatics and decision making, vol. 23, pp. 107, Jun. 2023, doi: 10.1186/s12911-023-02171-x.

[5] A. Imran, A. Nasir, M. Bilal, G. Sun, A. Alzahrani, and A. Almuhaimeed, "Skin cancer detection using combined decision of deep learners," IEEE Access, vol. 10, pp. 118198–118212, Nov. 2022, doi: 10.1109/ACCESS.2022.3220329.

[6] A. Echle, N. T. Rindtorff, T. J. Brinker, T. Luedde, A. T. Pearson, and J. N. Kather, "Deep learning in cancer pathology: a new generation of clinical biomarkers," British journal of cancer, vol. 12, pp. 686-696, Feb. 2021, doi: 10.1038/s41416-020-01122-x.

[7] G. Alfian et al., "Predicting breast cancer from risk factors using SVM and Extra-Trees-based Feature Selection Method," Computers, vol. 11, pp. 136, Sep. 2022, doi: 10.3390/computers11090136.

[8] Y. Lu et al., "A hybrid CNN-RNN approach for survival analysis in a Lung Cancer Screening study," Heliyon, vol. 9, pp. e18695, Aug. 2023, doi: 10.1016/j.heliyon. 2023.e18695.

[9] S. Alghunaim and H. H. Al-Baity, "On the scalability of machine-learning algorithms for breast cancer prediction in big data context," *IEEE Access*, vol. 7, pp. 91535–91546, Jun. 2019, doi: 10.1109/ACCESS.2019.2927080.

[10] A. Sahu, P. K. Das, and S. Meher, "High accuracy hybrid CNN classifiers for breast cancer detection using mammogram and ultrasound datasets," Biomed Signal Process Control, vol. 80, pp. 104292, Feb. 2023, doi: 10.1016/j.bspc.2022.104292.

[11] M. Javaid, A. Haleem, R. Pratap Singh, R. Suman, and S. Rab, "Significance of machine learning in healthcare: Features, pillars and applications," International Journal of Intelligent Networks, vol. 3, pp. 58–73, Jan. 2022, doi: 10.1016/j.ijin.2022.05.002.

[12] S. Bouhsissin, N. Sael, F. Benabbou, and A. Soultana, "Enhancing machine learning algorithm performance through feature selection for driver behavior classification," Indonesian Journal of Electrical Engineering and Computer Science, vol. 35, pp. 354, Jul. 2024, doi: 10.11591/ijeecs.v35.i1.pp354-365.

[13] X. Guo, M. Wei, H. Jiang, Z. Feng, and Y. Sun, "Data feature extraction methods based on deep learning," 2024 IEEE 4th International Conference on Electronic Technology, Communication and Information (ICETCI), IEEE, May 2024, pp. 1485–1489. doi: 10.1109/ICETCI61221.2024.10594224.

[14] M. J. Iqbal et al., "Clinical applications of artificial intelligence and machine learning in cancer diagnosis: looking into the future," Cancer Cell Int., vol. 21, pp. 270, May 2021, doi: 10.1186/s12935-021-01981-1.

[15] R. Chandragiri, D. B. Reddy, K. S. P. Badarla, A. H. Syed, K. Modepalli and K. Sadam, "A systematic review on machine learning techniques used for early detection of skin cancer," 2024 IEEE 13th International Conference on Communication Systems and Network Technologies (CSNT), IEEE, Jun. 2024, pp. 862–867, doi: 10.1109/CSNT.2024.140.

[16] M. Banda, E. K. Ngassam, and E. Mnkandla, "Enhancing classification and prediction through the application of hybrid machine learning models," 2024 IST-Africa Conference (IST-Africa), IEEE, May 2024, pp. 1–12. doi: 10.23919/IST-Africa63983.2024.10569590.

[17] S. Kusuma, S. G. Krishnan, K. Samreen, M. V. Ramana, and G. S. Prasad, "A hybrid deep learning approach for early detection and classification of lung cancer using the pelican optimization algorithm," 2024 International Conference on Distributed Computing and Optimization Techniques (ICDCOT), IEEE, Mar. 2024, pp. 1–6. doi: 10.1109/ICDCOT61034.2024.10515355.

[18] M. A. Zeller, Z. W. Arendsee, G. J. D. Smith, and T. K. Anderson, "classLog: Logistic regression for the classification of genetic sequences," Frontiers in Virology, vol. 3, pp. 1215012, Dec. 2023, doi: 10.3389/fviro.2023.1215012.

[19] K. Swanson, E. Wu, A. Zhang, A. A. Alizadeh, and J. Zou, "From patterns to patients: Advances in clinical machine learning for cancer diagnosis, prognosis, and treatment," Cell, vol. 8, pp. 186, Apr. 2023, doi: 10.1016/j.cell.2023.01.035.

[20] S. Molaei, S. Cirillo, and G. Solimando, "Cancer detection using a new hybrid method based on pattern recognition in micrornas combining particle swarm optimization algorithm and artificial neural network," Big Data and Cognitive Computing, vol. 8, pp. 33, Mar. 2024, doi: 10.3390/bdcc8030033.

[21] O. Khouadja, M. Jemai, and M. S. Naceur, "Improving cancer diagnosis with hybrid neural networks: A comprehensive narrative review," 2024 IEEE International Conference on Advanced Systems and Emergent Technologies (IC_ASET), IEEE, Apr. 2024, pp. 1–6. doi: 10.1109/IC_ASET61847.2024.10596187.

[22] A. Y. Saleh, C. K. Chin, V. Penshie, and H. R. H. Al-Absi, "Lung cancer medical images classification using hybrid cnn-svm," International Journal of Advances in Intelligent Informatics, vol. 7, pp. 151–162, Jul. 2021, doi: 10.26555/ijain.v7i2.317.

[23] S. Wankhade and S. Vigneshwari, "A novel hybrid deep learning method for early detection of lung cancer using neural networks," Healthcare Analytics, vol. 3, pp. 100195, Nov. 2023, doi: 10.1016/j.health.2023.100195.

[24] Y. S. Solanki et al., "A hybrid supervised machine learning classifier system for breast cancer prognosis using feature selection and data imbalance handling approaches," Electronics (Basel), vol. 10, pp. 699, Mar. 2021, doi: 10.3390/electronics10060699.

[25] X. Wang et al., "Characteristics of the cancer genome atlas cases relative to U.S. general population cancer cases," Br J Cancer, vol. 119, pp. 885–892, Oct. 2018, doi: 10.1038/s41416-018-0140-8.

[26] C. Fanconi, "Skin cancer: Malignant vs. Benign," ISIC archive Kaggle dataset. Accessed: Jan. 30, 2025. [Online]. Available: https://www.kaggle.com/datasets/fanconic/skin-cancer-malignant-vs-benign.

[27] Fazila Patel, "HybridML-SkinCancerDiagnosis, GitHub," 2024. Accessed: Jan. 31, 2025. [Online]. Available: https://github.com/patelfazila/HybridML-SkinCancerDiagnosis.

[28] L.-E. Pomme, R. Bourqui, R. Giot, and D. Auber, "Relative confusion matrix: Efficient comparison of decision models," 2022 26th International Conference Information Visualisation (IV), IEEE, Jul. 2022, pp. 98–103. doi: 10.1109/IV56949.2022.00025.

[29] A. Dongyao Jia, B. Zhengyi Li, and C. Chuanwang Zhang, "Detection of cervical cancer cells based on strong feature CNN-SVM network," Neurocomputing, vol. 411, pp. 112–127, Oct. 2020, doi: 10.1016/j.neucom.2020.06.006.

[30] P. E. Jebarani, N. Umadevi, H. Dang, and M. Pomplun, "A novel hybrid K-Means and GMM machine learning model for breast cancer detection," IEEE Access, vol. 9, pp. 146153–146162, Oct. 2021, doi: 10.1109/ACCESS.2021.3123425.

[31] B. Zheng, S. W. Yoon, and S. S. Lam, "Breast cancer diagnosis based on feature extraction using a hybrid of K-means and support vector machine algorithms," Expert Syst Appl, vol. 41, pp. 1476–1482, 2014, doi: 10.1016/j.eswa.2013.08.044.

[32] X. Wang et al., "Intelligent hybrid deep learning model for breast cancer detection," Electronics (Switzerland), vol. 11, Sep. 2022, doi: 10.3390/electronics11172767.

Comparative Case Study on Implementing Generative AI in Medical Practices to Ease Documentative Overburden: A Sociotechnical Systems Perspective

Sri Ramesh Eevani
Doctoral Candidate
LeBow College of Business, Drexel University
Philadelphia, USA
e-mail: ramesh.eevani@gmail.com

Rajiv Nag
Clinical Professor
LeBow College of Business, Drexel University
Philadelphia, USA
e-mail: rn362@drexel.edu

Abstract—**This paper presents a comparative case study of a live implementation of a Generative AI solution in 5 medical practices. Our findings shed new light on the impact of Generative AI on various aspects, such as social structures, roles, organizational processes, and technical systems of medical practices. It is well known now that the increasing documentation burden on physicians has led to medical errors, patient safety concerns, and physician burnout. This study investigates the adoption and implementation of a Generative AI-based clinical documentation technology in medical practices over 5 months. Our data included interviews, participant observations, process documentation and mapping, tracking social interactions, and analyzing textual user feedback data. The results enabled us to develop an implementation process framework that can be generalized across medical practices, categorizing changes into social, technical, organizational, and goals & outcomes. The implementation of Generative AI has led to both tangible and intangible benefits, including the creation of a new role of Scribe to provide human oversight of AI-generated clinical documentation. Resistance and apprehensions from practice staff have impacted implementation speed and decision-making. The study emphasizes the importance of considering social and organizational process changes in adopting new technologies and identifies role re-reforming and triadic co-creation as key concepts. Our process framework also includes an entrepreneur's and emerging technology product implementation team's co-creation experiences with the medical practices. Overall, this research provides a processual framework to capture the nuances of adopting and co-evolving an emergent and uncertain technology.**

Keywords-Physician Burnout; Documentation Overburden; Generative AI; Medical Practices; Clinical Documentation.

I. Introduction

Physicians are a crucial part of the healthcare delivery system, and their primary responsibility is to provide clinical treatment, medical advice, clinical documentation, and the best possible care to the patients. Although a physician's role can vary based on the peculiarities of the healthcare setting, in general, physicians provide services including preventive care measures, diagnosis of the ailment, referring other specialists, ordering medical tests, reviewing results, defining care plans, and explaining to the patients. There is an increased interest in making patients part of the decision-making process in a clinical setting, which is Shared Decision Making (SDM). SDM is a process where healthcare professionals and patients collaboratively make decisions based on reliable information, available options, and personal circumstances [3]. SDM requires physicians to spend very focused quality time assessing and discussing the clinical options and care plans. Physicians must also extensively document the clinical encounter details and the agreed-upon care plan with the patient.

Physicians are to document the patient encounter with detailed notes for dual purposes, record keeping of patient clinical notes, and for billing/insurance claim processing perspective. Clinical notes are crucial for government-regulated insurance plans such as Medicare and Medicaid. They can be audited by government agencies at any time, up to 7 years from the service date. For private commercial insurance plans, the payer can ask for detailed patient chart notes either as part of the claim adjudication process or to conduct periodic audits. Therefore, clinical documentation must be maintained in a timely and robust manner by the physicians to ensure effective delivery of patient care by other physicians coordinating the care and to avoid any auditing failures and regulatory penalties. The burden of clinical documentation responsibilities, however, limits the physicians' time to spend with the patients and provide high-quality clinical care. According to a survey conducted by Christino et al. [2] - A Nationwide Survey of Residents' Perceptions of Clinical Documentation Requirements and Patient Care, most physicians (92%) feel the documentation and regulatory obligations are excessive with 40% of the time for the documentation, and 12% with the patient at the bed side for clinical care.

The demand for extensive clinical documentation is increasing as regulations and insurance companies put a greater onus on physicians to document all aspects of patient care, treatment plans, procedural justifications, and any potential risks for clinical outcomes. The continual shift from clinical service and administrative tasks, such as clinical documentation, adds pressure to the physicians and contributes to their burnout. Extended office hours, continuous medical appointments, clinical tasks, administrative tasks, coordination between staff members, patient encounters, and other tasks lead to growing discontentment and dissatisfaction with current clinical documentation methods. This documentation overburden has contributed to medical errors, patient safety threats, lower quality of documentation and learning, and, ultimately, physician burnout [5]. Generative AI-based clinical documentation solutions can aid in ameliorating the current situation, thereby improving the productivity and performance of medical practices. That said, Generative-AI-based clinical documentation is still incipient, thereby requiring a deeper exploration of factors that impact the implementation of a novel yet proven technology.

The remainder of the paper is organized as follows: Section II provides the research methods we have used for this research study of cross-site comparison of medical practices adopting the emerging technology – Generative AI technology. In Section III, we explain the research study's results and describe the output from our analysis of the data we collected. Section IV further discusses the findings and the overarching process model we built and compares the change dimensions across the sites. In Section V, we provide the conclusion and opportunities for future research.

II. METHODS

Our study utilizes a qualitative research design with a hybrid approach that combines a sociotechnical systems approach with a comparative case study across five medical practices. We investigated the adoption and implementation of a Generative AI-based clinical documentation technology solution in 5 medical practices. We gained insights into how these practices differ when adopting uncertain and emergent technology. We used the cross-case comparative analysis method, first developed by Miles & Huberman [4]. It provides a structured approach to trace implementation processes within one practice site and then compare it with other sites (See Table 1). By employing this approach, we were able to develop a comprehensive understanding of the implementation processes, considering both the specific contextual realities of medical practices and the broader sociotechnical aspects at play.

A healthcare technology startup, Orci Care Inc., has been implementing its AI-based clinical documentation solution, built on Open AI's GPT 4.0 Large Language Model (LLM) version 2024-05-13 with human oversight service, at five medical practice sites. Four of the five medical practices are in upstate New York, and the fifth medical practice is in Atlanta, GA. These medical practices are physician-owned private practices, including Primary Care and Pediatrician specialties. We conducted 30 interviews with physicians, staff members, and members of the implementation teams from Orci Care Inc. across five medical practices where the product teams implemented the Generative AI technology solution. Through the implementation, we have collected the data over 5 months. In addition to the interviews, we also analyzed the data collected from field observations, pre-and post-implementation process documentation, and textual data collected through a product feedback form that the physicians in the medical practices filled out to track their views on using the Generative AI solution.

Implementing emerging technology such as Generative AI involves many unknown factors that might directly or indirectly influence the outcome of the implementation. Examples include whether to continue using the technology once the practice has made a commitment to it or when and how to decide that it is having an unfavorable impact on the practice and terminating its use. We explored various factors in adopting emerging technology in various medical practices to compare multiple dimensions. We also investigated the changes that occur during and post-implementation from the perspectives of the technology itself, the process of introducing it, social factors, the goals of the practice and the outcomes of the technology's use.

In addition to the interviews, we also analyzed the data collected from the notes on the field study, documentation about the pre-and post-implementation process, social interactions, and responses to a feedback survey. The diverse data sources will make the qualitative analysis more solid and incorporate various viewpoints. Grounded theory methods provide guidelines for collecting and analyzing the data systematically and making sense of the data while building the

TABLE I. DIMENSIONS OF CHANGE WITH GENERATIVE AI ADOPTION

	Site-1	Site-2	Site-3	Site-4	Site-5
Social Change					
Substitutive Apprehending	Low	Low	X	High	Med
Adoption Priming	++	+	N/A	(-)	++
Pre-AI Interactional Dissonance	Med	Low	Low	Low	Med
Post-AI Interactional Enrichment	++	+	++	X	+
Competence Scaffolding	Med	Med	Low	High	Med
Technical Change					
Technology Deficiency	Med	High	Low	Low	Med
Implementative Co-maturing	High	Med	Med	High	High
Continual Technology Tinkering	Med	Med	Low	High	Med
Triadic Co-creating	High	Med	Low	High	Med
Oversight Enforcing	High	Med	Low	High	High
Organizational Change					
Administrative Burdening	High	Low	Med	Med	High
Solution Exploring	High	Low	Med	High	High
Role Reforming	2	1	2	1	2
Perceived Change Load	Low	High	Low	Med	Med
Process Simplifying	++	+	++	+	++
Calibrated Onboarding	High	Med	Low	Med	High
Goals & Outcome					
Prospecting Trailblazers	High	Low	Med	Med	High
Affirmative Verifying	High	Low	High	Med	High
Comparative Verifying	Low	Low	Med	High	Low
Performative Verifying	High	Low	Med	Med	High
Realized Beneficence	++	+	++	X	++
Potential Beneficence	+	+	X	X	+
User Trust Accreting	++	+	++	+	++
Ultimate Outcome					
Continued AI Use	Positive	Negative	Positive	Neutral	Positive

Numbers (0 – 2) = Number of new job roles
++ = Significant positive change
+ = Noticeable positive change
N/A = Not Applicable/Available
(-) = Noticeable negative change
X = No noticeable change
Low = Low change/level
Med = Medium change/level
High = High change/level

theoretical frameworks [1]. Grounded theory allows the researcher to identify the patterns in the data and build the theoretical concepts from the data rather than beginning with a set of hypotheses to prove [1]. While my research is not entirely based on grounded theory, we leveraged the concepts of data collection and the grouping of the data to construct my framework.

III. Results

The analysis of data collected over 5 months on Generative AI implementation across five medical practices led to the discernment of a process framework (see Figure 1). The goal was to conceptualize the findings across a broader range of Generative AI-based technologies for adoption in medical practices. We created initial codes of the information from the 30 interviews using the online program Delve. Next, following the grounded theory, we created focused codes that identified the overarching concepts from the initial codes and first-order categories. The focused codes represent the generic concepts that can be applied beyond the specific scope of this study to the adoption of any emerging technology in medical practices. We used the techniques of Miles and Huberman [4] to visualize various process elements and concepts and document the resulting displays. These visuals helped develop a processual map of implementing the emerging technology. It should be useful for further research and helpful for those seeking to introduce future emerging technology into medical practices. We have developed the overarching process model for Site-1 and enhanced it to incorporate the process model from the other four sites, resulting in a generalized process model that cuts across all five medical practice sites. Our analysis revealed that implementing the Generative AI technology created different adoption experiences across the practices. The changes observed across the sites are categorized as social changes, technical changes, organizational changes, and goals & outcomes. This approach is apt and suitable for this research study as the emerging technology adoption across multiple sites.

The comparative case study across the five medical practice sites provided insights into the adoption experiences of the physicians and the elements of their ability to adopt the change, co-creative savviness, and patience levels to sustain initial disruptions. Adoption priming is preparing and supporting the end users through the initial adoption stages, sustaining the disruption with minimal impact, and assisting in achieving long-term benefits of the emerging technology implementation. The sites have experienced different levels of adoption priming based on physicians' technical savviness, staff reluctance levels to support the change, and job security concerns. Etc.

In a typical matured and stable technology implementation and adoption process, the main factors of the adoption include the technical systems, implementation complexity, and end-user readiness. However, in emerging technology implementations, additional consideration is given to the ambiguity of the end user in trusting the technology and acceptance levels of the disruptions during the implementation and stabilization phases. As Generative AI technology can potentially challenge and substitute traditional human roles in creating such content, the technology is inciting job security fears. We found that job security concerns with the emerging technology were present in some sites significantly more than the others. We also found that some practices created a new job role for Scribe as a human oversight of the AI-based emerging technology.

The overarching process model, as shown in Figure 1, describes the end-to-end view of emerging technology adoption at medical practices from a sociotechnical systems perspective. The horizontal view shows the progression of emerging technology adoption phases. The pre-AI phase describes process elements experienced at the sites before the AI adoption. It shows the existence of interactional dissonance between physicians and patients as physicians experience administrative overburden and are distracted from taking notes while treating the patients. The pre-AI Product Discovery Phase includes the processual elements of exploring and evaluating solution options by the medical practices to solve the administrative burden, engaging the Generative AI-based product team, and learning more about the product. In this phase, it is observed in a few sites that substitutive apprehensions from the practice staff with the fear of job security exist. The AI Implementation phase includes the process elements involving how the product team and medical practice collaborate with the technology implementation and initial adoption disruptions. The collaboration between the product team, the physician, and the medical practice staff generates co-creation, adoption priming, competence scaffolding, and continual technology tinkering. The product team continues to enhance the product with the physician and medical practice staff's feedback, calibrating the onboarding process and co-maturing the product implementation. Physicians enforcing the human oversight of the AI output for clinical documentation resulted in role re-forming and generating new roles at the medical practice as "Scribe". Oversight enforcement and process simplification with the technology adoption promoted user trust accretion on the product & the emerging technology as physicians continue to verify the technology and realize the tangible and intangible benefits of the AI technology and the adoption outcome of the perpetuating AI use. It is observed across the sites that adopting AI technology for clinical documentation resulted in patient interactional enrichment.

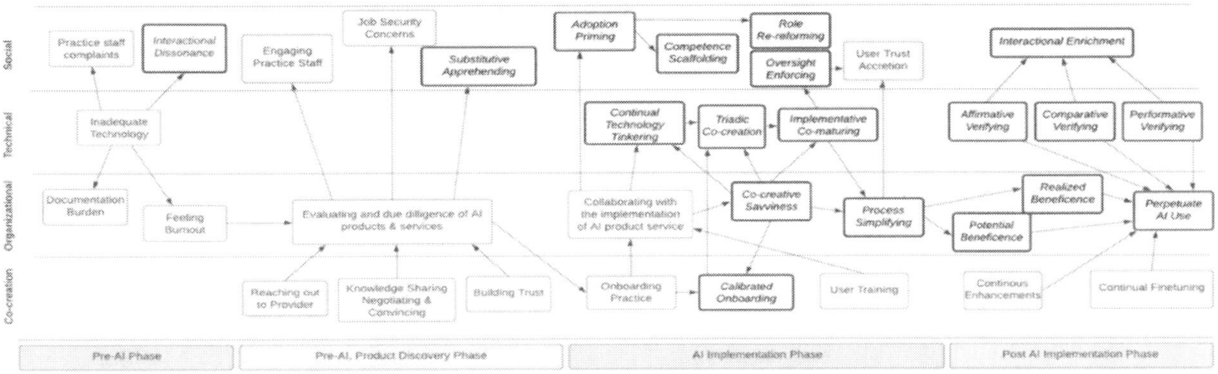

Figure 1. Overarching Process Model (Across All Sites)

IV. Discussion

Our study finds that implementing a Generative-AI-based technological solution for clinical documentation has led to several potential intangible and tangible benefits for physicians and medical practices. For instance, the implementation has seen the emergence of a new role in medical practices, that of a *Scribe* as human oversight. The scribe's role is to maintain critical oversight and conduct careful quality control of the clinical documentation generated by AI by verifying and curating it for the physician's consumption. We also find evidence of role-reforming in the medical practice in coordinating the clinical documentation activities between physicians, scribes, clinicians, and other practice staff. Some sites encountered varying levels of resistance from the practice staff with substitutive threats and apprehensions that significantly impacted implementation speed and subsequent AI use decisions. Furthermore, we found that professionals in charge of implementing emerging technologies need to consider the physicians' adoption ambivalence and substitutive apprehensions of the practice staff and can handle it effectively through implementative co-maturing.

Technology startup teams of emerging technology such as Generative AI have a very tricky situation to handle, as the product team needs to continue to monitor and adopt the underlying unmatured technology such as Artificial Intelligence (AI) while evolving the product and services to implement and provide tangible benefits to the customers. Notably, our study shows the critical importance of triadic co-creation as an element of the implementation process. This research is significantly different from the implementation of traditional technologies in the sense that in the context of Generative-AI, the co-creation process involves mutual adaptation between the technology implementation team, the user (personnel in the medical practices), and the AI-based technology that can autonomously learn and modify itself. We found that triadic co-creation occurs when product teams work closely with early customers who are equally motivated to achieve tangible and intangible benefits with the emerging technology.

V. Conclusion and Future Work

This research offers a novel multi-dimensional perspective on adopting Generative AI in medical practices, focusing on a sociotechnical system approach. The research included the comparative qualitative study of Generative AI adoption at 5 medical practices. It emphasizes the role and the importance of considering social, organizational process changes, and technical systems when adopting new technologies. The study provided an implementation process of emerging technical adoption at medical practices from the perspectives of social, organizational, and technical, and goals & outcomes categories. The research found that Generative AI adoption at medical practices resulted in tangible and intangible benefits to the practice and in most of the cases a new role of Scribe has evolved. It also found that resistance and apprehensions from the practice staff has resulted in adoption speed and the overall outcome. The study identifies role re-reforming and triadic co-creation as the key elements in implementing an emerging technology at medical practices.

References

[1] Charmaz, K. 2006. *Constructing Grounded Theory: A Practical Guide Through Qualitative Analysis*. London: Sage Publications.

[2] Christino, M. A., Matson, A. P., Fischer, S. A., Reinert, S. E., Digiovanni, C. W., & Fadale, P. D. (2013). Paperwork Versus Patient Care: A Nationwide Survey of Residents' Perceptions of Clinical Documentation Requirements and Patient Care. *Journal of Graduate Medical Education*, *5*(4), 600–604. https://doi.org/10.4300/JGME-D-12-00377.1

[3] Driever, E. M., Stiggelbout, A. M., & Brand, P. L. P. (2020). Shared decision making: Physicians' preferred role, usual role and their perception of its key components. *Patient Education and Counseling*, *103*(1), 77–82. https://doi.org/10.1016/j.pec.2019.08.004

[4] Miles, M., & Huberman, A. M. (1994). Qualitative data analysis: An expanded sourcebook (2nd ed.). Thousand Oaks, CA: Sage.

[5] Moy, A. J., Schwartz, J. M., Elias, J., Cato, K. D., Small, D. S., & Rossetti, S. C. (2021). Measurement of clinical documentation burden among physicians and nurses using electronic health records: A scoping review. *Journal of the American Medical Informatics Association, 28*(5), 998–1008. https://doi.org/10.1093/jamia/ocaa325

Personalized Automated Blood Glucose Forecasting for Type-1 Diabetes Using Machine Learning Algorithms

Avijay Sen,
Franklin High School
Elk Grove, California, United States
avijay.sen12@gmail.com

Dr. Sindhu Ghanta
AIClub Research Institute
Mountain View, CA, USA
sindhu@aiclub.world

Pallavi Bajpai
AIClub Research Institute
Mountain View, CA, USA
pallavi.bajpai@aiclub.world

***Abstract*—Type-1 Diabetes Mellitus (T1DM) is a chronic condition characterized by the pancreas's inability to produce insulin, requiring continuous monitoring and management of blood glucose levels. Accurate prediction of blood glucose levels can significantly improve patient outcomes by reducing hypo- and hyperglycemic events. This study develops a personalized automated blood glucose forecasting system leveraging the past blood glucose levels and insulin pump data. Utilizing the publicly available Diatrend dataset, encompassing thirty-one days of data for five subjects, we evaluated three machine learning algorithms: K-Nearest Neighbors (KNN), Random Forest (RF), and Multilayer Perceptron (MLP). After hyper-parameter tuning, the performance of each algorithm was assessed using Root Mean Squared Error (RMSE), Mean Squared Error (MSE), and the coefficient of determination (R^2), with a particular emphasis on RMSE. The Random Forest model demonstrated superior performance, achieving a test RMSE range of $14.98-23.62$ across all subjects. This research highlights the efficacy of supervised machine learning algorithms in predicting blood glucose levels over one-hour intervals for T1DM patients, underscoring the potential of personalized machine learning models to improve diabetes management.**

Keywords- blood glucose prediction; Type-1 Diabetes Mellitus; insulin delivery system

I. Introduction

Type 1 Diabetes (T1DM) is a chronic condition where the pancreas fails to produce insulin, the hormone needed to control blood sugar levels. People with T1DM face challenges in managing blood sugar, which can be too low (hypoglycemia) or too high (hyperglycemia). Low sugar levels, below 70 mg/dL, can cause symptoms like sweating, hunger, and even serious issues like seizures or coma [1], [2]. High sugar levels, over 140 mg/dL, may lead to problems in the eyes, heart, and nerves [3], [4]. Managing these fluctuations requires careful insulin use, which can be challenging and risky [5].

The eight leading cause of death globally is diabetes [6]. The number of deaths has been increasing since the start of the 21st century [7]. The increasing trend approximates that there will be 13.5-17.4 million people suffering from T1DM by 2040 [8]. Majority of the deaths occur before 70 and are due to high glucose levels [9].

A development of a sophisticated insulin delivery method that combines Continuous Glucose Monitoring (CGM), which utilizes the subcutaneous interstitial fluid to measure glucose levels and insulin pumps which use glycemic data from the monitors to provide temporary insulin formulas like basal or bolus to maintain glucose levels. The device asks the patient for information on physical activity, insulin bolus dosage, meal sizes and carbohydrate content, among other things, in order to obtain more accurate assessments [10].

To further enhance the capabilities of CGMs, Machine Learning (ML) offers a promising avenue. ML can perform human-like tasks through learning from data and being able to adapt to unseen data. There are various types of ML algorithms, such as Supervised, Unsupervised, Semi-supervised, and Reinforcement learning [11]. Supervised learning is typically the task of ML to learn a function that maps an input to an output based on sample input-output pairs [12]. It uses labeled training data and a collection of training examples to infer a function. Supervised learning is carried out when certain goals are identified to be accomplished from a certain set of inputs [13]. There are two different types of common supervised tasks which include "classification" that separates the data or "regression" that fits the data [12]. For the purpose of this study, regression was used consisting of different algorithms [14].

The integration of technology in diabetes management has led to significant advances in the prediction and control of blood glucose levels [15]. CGMs combined with insulin pumps, forms the backbone of artificial pancreas systems, which automate insulin delivery to maintain optimal glucose levels [16]. These closed-loop systems have shown promise in reducing the burden of daily diabetes management and improving overall quality of life for patients [17]. Studies have demonstrated that such systems can significantly improve glycemic control, reduce HbA1c levels, and mitigate the risks associated with long-term diabetes complications [18]. The continuous evolution of these technologies and their integration with machine learning algorithms hold the potential to transform diabetes care, making it more precise, personalized, and effective [19].

Our paper focuses on evaluating closed-loop insulin delivery systems, known as artificial pancreas systems, for their effectiveness and safety in managing T1DM. By analyzing CGM data, we developed a method to fine-tune insulin rates using various ML models. Our personalized approach using the Diatrend dataset demonstrates the strength and flexibility of these models for individual patient needs.

The paper is structured as follows: The Introduction discusses the challenges of T1DM and the role of ML in improving insulin systems. The Related Work section reviews existing models and their limitations. Materials and Methods explain our dataset, data preparation, and methodology. Results provide an analysis of model performance. Discussion interprets the findings, comparing them with existing methods. Future Work & Limitations suggest improvements and study constraints. The Conclusion summarizes our contributions and highlights the importance of personalized systems in diabetes care.

II. Related Work

Predicting blood glucose levels in patients with T1DM has been the focus of numerous studies employing a variety of machine learning algorithms and models. Machine learning plays a crucial role in predicting blood glucose levels by analyzing vast amounts of data to identify patterns and trends that are not easily discernible by traditional methods. This allows for more accurate and personalized predictions, ultimately improving diabetes management and reducing the incidence of hypo- and hyperglycemic events. Prior research has demonstrated the potential of different methods, yet each approach has limitations that impact the predictability and efficiency of the models.

The emergence of CGMs has introduced different methodologies aimed at forecasting glucose levels. There have been advancements in creating physical models and/or data-driven observational models that attempt to predict glucose levels of patients [20]. A few models that have been used are Proportional-Integral-Derivative (PID) Controllers [21], Artificial Neural Networks (ANNs) [22], Recurrent Neural Networks (RNNs) [23], Long Short-Term Memory (LSTM) Networks [24], Support Vector Machines (SVM) [25], [26], Fuzzy Logic Systems [27], and RFs [28]. Recently, neural network based models are gaining popularity: the use of dilated recurrent neural networks (DRNNs), which have shown promise in improving prediction accuracy by handling sequential data more effectively and overcoming issues like gradient vanishing [29]. Additionally, transfer learning approaches, where models are initially trained on a generalized dataset and then fine-tuned with individual patient data, have demonstrated enhanced prediction accuracy for specific subjects [30].

One notable study, titled "A Machine Learning Approach to Predicting Blood Glucose Levels for Diabetes Management" implemented the Support Vector Regression (SVR) algorithm alongside a physiological model characterized by three compartments: meal absorption dynamics, insulin dynamics, and glucose dynamics. The researchers utilized a small sample size of five T1DM patients to pull different parameters including carbohydrate intake, rapid-acting insulin, bolus and basal rate, body mass, and insulin sensitivity (IS) [31].

Similarly, another research paper compared the efficacy of LSTM networks and Temporal Convolutional Networks (TCNs) for blood glucose level prediction [32]. This study also explored various classification algorithms, including SVM, Naive Bayes, and Decision Tree for comparison. The results indicated that there was little benefit to employing TCN or LSTM over conventional models, pointing to a potential application gap for these cutting-edge neural networks. This emphasizes the necessity of more research to determine the circumstances in which these models could provide meaningful advantages.

Further research evaluating the accuracy of SVM, Naive Bayes, and Decision Tree algorithms in diabetes classification were conducted using the Pima Indian Diabetes Database [33]. One weakness of the dataset was its homogeneity—all of the patients were of the same race. This limited the results' applicability to more diverse populations with a range of genetic and lifestyle backgrounds.

In order to overcome issues like missing data, research has also been done using RNN algorithms to predict blood glucose levels [34]. The study focused on improving prediction accuracy by utilizing the temporal dependencies in CGM data. The existence of missing data, however, presented a serious problem and might have an effect on the model's predictability and accuracy. Developing efficient methods to deal with missing data is essential to enhancing RNN models' resilience in practical applications.

Interestingly, researchers have proposed a hybrid approach combining SVM and Neural Networks (NN) to improve blood glucose level predictions. This method demonstrates enhanced accuracy in glucose forecasting, particularly in reducing prediction errors compared to traditional models [35]. However, the study relies on a relatively small dataset, which may affect the generalizability of the model to broader populations.

Building on these efforts, we adopt a different approach by utilizing the same dataset as prior studies but with distinct model choices and methodology. While deep learning models like LSTM and Encoder-Decoder are commonly used for time-series predictions, as highlighted in "Deep Learning-Based Glucose Prediction Models: A Guide for Practitioners" [36], we opt for simpler machine learning techniques such as KNN, RF, and MLP in order to easily integrate into healthcare systems. Additionally, we focus on hyperparameter optimization for individual subjects rather than complex training strategies like personalized or fine-tuning methods. This allows us to prioritize model simplicity and interpretability while still leveraging the same data.

In contrast to other approaches, our methodology involves the use of three distinct algorithms: KNN, RF, and MLP, to provide personalized solutions for each patient. The dataset we used includes a diverse group of subjects with varied characteristics, such as differences in sex and race, enhancing the representativeness and predictability of our results. By conducting hyperparameter tuning and training multiple models, we selected the best-performing model to ensure the robustness and accuracy of our findings, setting our research apart from previous studies.

This body of related work highlights the ongoing efforts and challenges in predicting blood glucose levels in T1DM patients. Each study contributes uniquely to the field, offering insights and advancements while highlighting areas for further investigation and improvement.

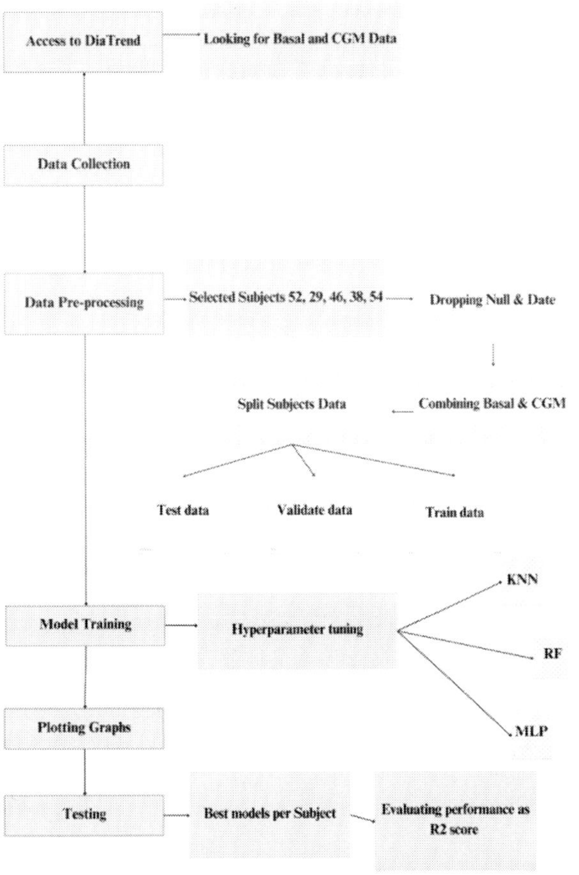

Figure 1. Flowchart of the process.

III. Materials and Methods

This section outlines the materials and methods employed in this study, providing a detailed framework for the research process.

A. Dataset

We utilized the Diatrend dataset [37], which offers extensive continuous data from wearable medical devices. This includes 8,220 days of insulin pump data and 27,561 days of CGM data from 54 diabetic patients. For our analysis, we selected five subjects out of 17 subjects from this dataset who had comprehensive CGM and basal insulin readings available to ensure data completeness.

B. Data Pre-processing

The dataset was initially filtered to identify patients with both CGM and basal insulin readings. To prepare the data for analytical and statistical models, several pre-processing steps were undertaken to ensure data quality and completeness.

Basal insulin data entries, which include fields for "date," "rate" (units per hour), and "duration" (milliseconds), were adjusted to ensure that no single duration exceeded 5 minutes (300,000 milliseconds). Any entries with duration longer than 5 minutes were split into multiple 5-minute segments, and the corresponding timestamps were updated accordingly. This adjustment facilitated accurate alignment with CGM data, ensuring consistent time intervals.

The modified basal insulin data was integrated with the CGM data to create a unified dataset. For each CGM timestamp, the corresponding basal insulin data was merged. If multiple basal insulin entries existed within the interval between two CGM readings, the basal entry that either matched or immediately followed the CGM timestamp that was selected.

Missing values in the CGM data for columns like "mg/dl" (glucose concentration) column were addressed using forward-fill imputation. This method replaces missing values with the last observed value, which is appropriate for maintaining the continuity of time-series data. Both CGM and basal insulin datasets were sorted by date to preserve their temporal sequencing.

To capture both glucose trends and insulin delivery patterns over time, the following features were calculated using a rolling window of 12 data points (equivalent to 1 hour if readings are taken every 5 minutes): Glucose Mean (glucose_mean): The mean glucose level over the window. Glucose Standard Deviation (glucose_std): The standard deviation of glucose levels over the window. Weighted Basal Infusion (basal_infusion): This feature was calculated as the sum of the product of "duration" and "rate" divided by the sum of "duration" over the window, representing the average basal insulin delivery rate weighted by duration.

These features provided a comprehensive view of glucose dynamics and insulin administration, which are critical for predictive modeling in diabetes management. After feature extraction, the dataset for each data point included the following features:

- date: Timestamp of the CGM reading.
- glucose_mean: Mean glucose level over the past hour.
- glucose_std: Standard deviation of glucose levels over the past hour.
- basal_infusion: Weighted average basal insulin infusion rate over the past hour.
- mg/dl: Current glucose reading.

After this, we divided the dataset in the order of time to preserve the temporal order and prevent any mixing of future and past data, hence randomization was not an option. By doing this, we made sure that the model learned from earlier data and was tested on later data, similar to real-world prediction situations, maintaining the quality of our time-based analysis.

- Training Set (70%): The earliest 70% of the data points, used to train the model.
- Validation Set (15%): The subsequent 15% of data points, used for hyper-parameter tuning.
- Test Set (15%): The latest 15% of data points, used to evaluate the model's performance on unseen data.

C. Methodology

For our analysis, we chose KNN, RF, and MLP regression models because of their proven effectiveness in both time-series prediction and glucose level forecasting:

- KNN: Valued for its straightforward approach and ability to capture local data patterns, KNN has successfully been used in glucose prediction, yielding satisfactory outcomes [38].
- RF: This ensemble learning technique improves predictive accuracy and mitigates over fitting. RF models are known for their robust performance in analyzing medical data and offer feature importance metrics, enhancing interpretability [39].
- MLP: As a type of neural network, MLP excels at modeling complex, non-linear relationships, making it highly appropriate for glucose prediction where such intricate patterns are present [40].

In particular, the feature importance scores provided by RF models significantly boost interpretability, which is essential in personalized medicine. Our choices emphasize a balance between achieving high predictive performance and maintaining model interpretability.

The dataset consists of five subjects. The dataset comprises data from five distinct subjects. For each subject, we developed a unique model using data specific to that individual since subject's timestamps were different for all subject's readings. Each of the three algorithms was applied separately to the data from each subject, allowing us to conduct thorough experiments tailored to each subject's dataset.

We conducted hyper-parameter tuning for each model to enhance performance. For KNN, we tested using between 1 and 16 neighbors to find the right balance for understanding both small and large patterns in the data. With RF, we experimented with using between 10 and 100 decision trees and adjusted their depth from 1 to 7 to avoid making the model too complex or too simple. For the MLP, we varied the starting learning rates between 0.00001 and 0.05 and adjusted the number of iterations from 10 to 100 to see how these factors affected the model's learning and improvement speed. Each configuration's performance was assessed using RMSE and R^2 on the validation set. This tuning process was crucial for ensuring generalization and avoiding over fitting. The optimal hyper-parameters differed across subjects, reflecting the unique glucose dynamics of each individual [41]. Furthermore, the best-performing model was employed to evaluate its performance by applying it to the test data of all subjects.

Following an extensive hyper-parameter tuning phase, the models that exhibited the best performance based on validation metrics were selected. These models were then rigorously tested on each of the five subjects' test data to evaluate their reliability. This evaluation involved calculating three key performance metrics: MSE, RMSE, and the R^2. These steps ensured an assessment of the model's predictive capabilities, providing insights into their performance on data that was not used during training and hyper-parameter tuning. Refer to Figure 1 for a visual representation of the Materials and Methods processes.

IV. Results

In this section, we present the performance of three machine learning models— KNN, RF, and MLP— across five different subjects, using the RMSE and R^2 (coefficient of determination) score as the key metric.

Graphs illustrating model performance metrics for each algorithm and subject using validation data are shown in Figure 2. Specifically, we plotted RMSE against the K values for KNN models, RMSE against the number of estimators (n_estimators) for various max_depth configurations in RF models, and RMSE against the number of iterations for different learning rates in MLP regression models.

The graph displays an RMSE range of 14.98 to 23.62 mg/dL. While this is relatively high, it falls within acceptable limits for glucose prediction models. Given that glucose levels can vary significantly and rapid fluctuations are common in Type 1 Diabetes patients, clinical guidelines typically consider deviations within ±30 mg/dL to be acceptable. Therefore, our model's errors are within a clinically relevant range [42].

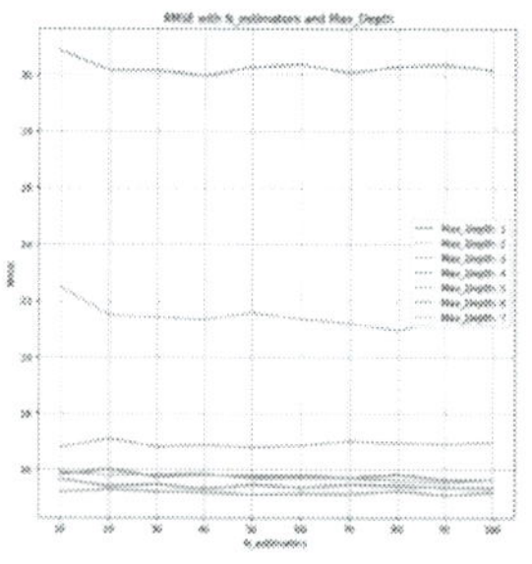

(a) Subject 52 (RF)

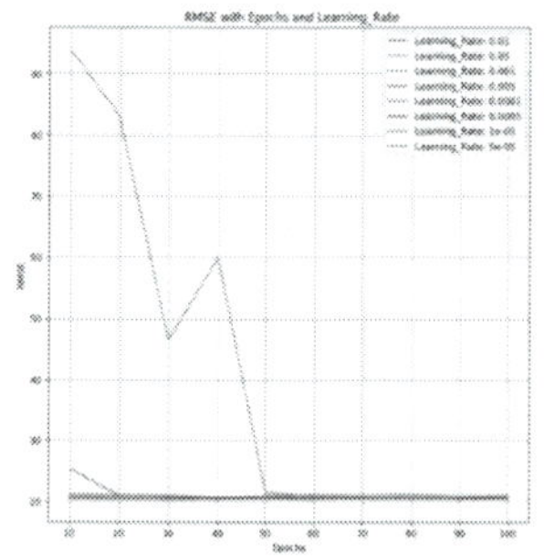

(b) Subject 29 (MLP)

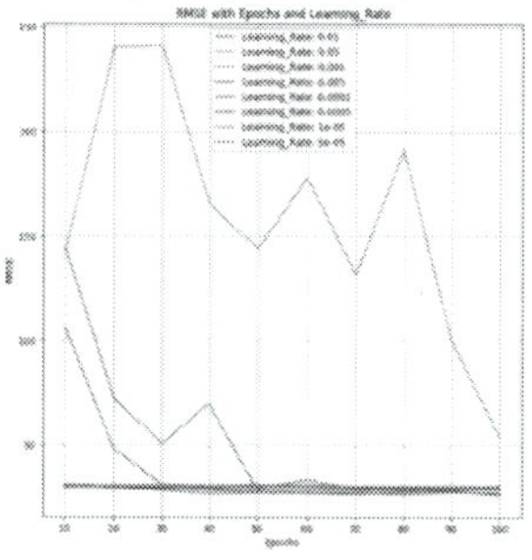

(c) Subject 46 (MLP)

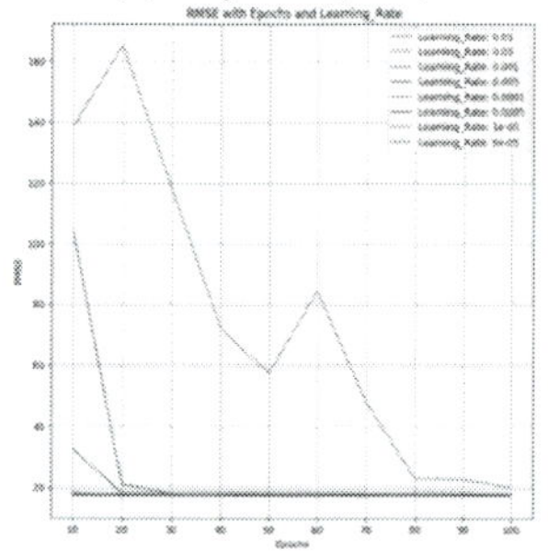

(d) Subject 38 (MLP)

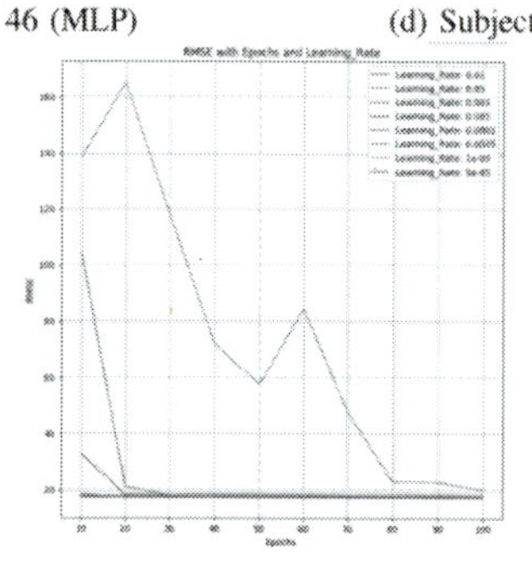

(e) Subject 54 (RF)

Figure 2. Subject's graph of their best performing model

Results of the best model obtained from hyper-parameter tuning on the validation and test datasets is shown in Tables I and II.

TABLE I. TRAINING RESULTS FOR DIFFERENT SUBJECTS AND MODELS

ID	KNN	RF	MLP
52	MSE: 252.947 RMSE: 15.904 R2 Score: 0.912	MSE: 227.535 RMSE: 15.084 R2 Score: 0.921	MSE: 317.137 RMSE: 17.808 R2 Score: 0.890
29	MSE: 438.806 RMSE: 20.947 R2 Score: 0.857	MSE: 425.273 RMSE: 20.622 R2 Score: 0.861	MSE: 420.411 RMSE: 20.503 R2 Score: 0.863
46	MSE: 814.730 RMSE: 28.543 R2 Score: 0.880	MSE: 717.231 RMSE: 26.781 R2 Score: 0.895	MSE: 820.608 RMSE: 28.646 R2 Score: 0.879
38	MSE: 317.209 RMSE: 17.810 R2 Score: 0.866	MSE: 310.727 RMSE: 17.627 R2 Score: 0.869	MSE: 301.532 RMSE: 17.364 R2 Score: 0.873
54	MSE: 342.127 RMSE: 18.496 R2 Score: 0.772	MSE: 299.137 RMSE: 17.295 R2 Score: 0.800	MSE: 375.030 RMSE: 19.365 R2 Score: 0.750

TABLE II. TESTING RESULTS FOR DIFFERENT SUBJECTS AND MODELS

ID	KNN	RF	MLP
52	MSE: 314.087 RMSE: 17.722 R2 Score: 0.926	MSE: 305.725 RMSE: 17.484 R2 Score: 0.928	MSE: 378.007 RMSE: 19.442 R2 Score: 0.911
29	MSE: 414.655 RMSE: 20.363 R2 Score: 0.880	MSE: 391.740 RMSE: 19.792 R2 Score: 0.886	MSE: 385.436 RMSE: 19.632 R2 Score: 0.888
46	MSE: 615.205 RMSE: 24.803 R2 Score: 0.922	MSE: 558.373 RMSE: 23.629 R2 Score: 0.929	MSE: 546.354 RMSE: 23.374 R2 Score: 0.931
38	MSE: 352.870 RMSE: 18.784 R2 Score: 0.800	MSE: 340.414 RMSE: 18.450 R2 Score: 0.807	MSE: 330.102 RMSE: 18.168 R2 Score: 0.813
54	MSE: 235.849 RMSE: 15.357 R2 Score: 0.789	MSE: 224.320 RMSE: 14.977 R2 Score: 0.800	MSE: 293.482 RMSE: 17.131 R2 Score: 0.738

For the training data, the Random Forest model achieved the highest R^2 score of 0.921 for Subject 52, demonstrating better predictive ability compared to the KNN model with an R^2 score of 0.912 and the MLP model with an R^2 score of 0.890. For Subject 29, the MLP model emerged as the best performer with an R^2 score of 0.863, slightly outperforming the RF and KNN models, which had R^2 scores of 0.861 and 0.857, respectively. In the case of Subject 46, the RF model again showed the highest predictive ability with an R^2 score of 0.895, while KNN and MLP had similar performances, with R^2 scores of 0.880 and 0.879, respectively. Additionally, for Subject 38 the MLP model achieved the highest R^2 score of 0.873, indicating better performance than both the RF and KNN models, which had R^2 scores of 0.869 and 0.866, respectively. Finally, for Subject 54, the RF model outperformed the other models with an R^2 score of 0.800, followed by the KNN model with an R^2 score of 0.772, and the MLP model with the lowest performance at an R^2 score of 0.750.

Overall, the RF model consistently executed the highest R^2 scores across the majority of subjects, indicating strong predictive performance. Specifically, the RF model had the highest R^2 scores for Subject 52 (0.921), Subject 46 (0.895), and Subject 54 (0.800). The MLP model performed the best for Subject 29 (0.863) and Subject 38 (0.873). While the KNN model showed strong performance, it did not outperform the RF or MLP models in any subject. These findings imply that the MLP and KNN models are closely followed by the RF model, which is the most reliable option for precise predictions across this dataset.

When evaluating the models on the test data, the RF model obtained the highest R^2 score of 0.928 for Subject 52, closely followed by the KNN model with an R^2 score of 0.926. The MLP model had a slightly lower R^2 score of 0.911. This indicates that both RF and KNN models performed similarly well, slightly outperforming the MLP model for this subject. In the case of Subject 29, the MLP model emerged as the best performer with an R^2 score of 0.888. The RF model also performed well, achieving an R^2 score of 0.886, while the KNN model had a slightly lower R^2 score of 0.880. The differences in performance were minimal, suggesting that all three models were effective for this subject, with the MLP model having a slight edge. For Subject 46, the MLP model demonstrated the highest predictive performance with an R^2 score of 0.931, followed by the Random Forest model with an R^2 score of 0.929. The KNN model also performed strongly with an R^2 score of 0.922, but was slightly surpassed by the other two models. For Subject 38, the MLP model again accomplished the highest R^2 score of 0.813, indicating fitter performance than both the RF model (R^2 score of 0.807) and the KNN model (R^2 score of 0.800). All three models performed well, but the MLP model was the best among them for this subject. Finally, for Subject 54, the RF model exceeded the other models with an R^2 score of 0.800. The KNN model followed with an R^2 score of 0.789, while the MLP model had the lowest performance with an R^2 score of 0.738.

In summary, the performance of each model varied across different subjects, but overall, the RF and MLP models frequently demonstrated superior predictive capabilities. Specifically, the Random Forest model achieved the highest R^2 scores for Subject 52 (0.928) and Subject 54 (0.800), while the MLP model led for Subject 29 (0.888), Subject 46 (0.931), and Subject 38 (0.813). The KNN model showed strong performance but was generally outperformed by the RF and MLP models. These results underscore the value of using multiple models to identify the most effective predictive approach for different datasets.

We subsequently selected the best-performing model, identified by its lowest RMSE score of 14.977, as the most effective approach. To assess the robustness and generalizability of this model, we applied it to the test data of all subjects, evaluating its performance across the entire dataset. This approach allowed us to determine whether the optimized model could maintain its accuracy and reliability when exposed to diverse subject-specific data. Table III presents the model's applicability and performance metrics for each subject.

TABLE III. TESTING RESULTS FOR SUBJECTS ON BEST MODEL

ID	RMSE
52	RMSE: 31.300
29	RMSE: 22.552
46	RMSE: 43.736
38	RMSE: 18.716
54	RMSE: 14.977

V. Discussion

In this study, we developed models to predict blood glucose levels of patients using machine learning algorithms. We tested three different algorithms: KNN, RF, and MLP on five different subject's datasets from the Diatrend dataset. The process for each subject's dataset consisted of training, validation, and testing of the models. The performance of these models was evaluated based on three metrics namely, R^2, MSE, and RMSE. However, for the scope of this study, we narrowed down our analysis for determining best performance to rely more heavily on R^2 and RMSE.

The model hyper-parameters chosen for each patient dataset impacted the RMSE and R^2 values differently for each subject. KNN algorithm did not yield a high performance in any of the subjects.

In case of random forest, for subjects 52 and 54, moderate values of hyper-parameters yielded the highest performance. For Subject 52, the results indicate that increasing the maximum depth and number of estimators generally improves model performance up to a certain point. The lowest MSE and RMSE values are observed at a maximum depth of 5 and 90 estimators, with an MSE of 15.124 and RMSE of 15.085. However, further increases in these hyper-parameters do not lead to substantial improvements and, in some cases, result in slightly worse performance. The results for Subject 54 show a slightly different set of trends. Here, the learning rate and the number of epochs play a crucial role in model performance. It is clear that excessively high or low learning rates lead to poor performance, as evidenced by the extremely high MSE values for learning rates of 0.01 and 0.00001. The most optimal performance is observed at a learning rate of 0.05 with 90 epochs, yielding an MSE of 17.055. While moderate values of hyper-parameters tend to yield better performance generally, the specific sensitivity varies between subjects.

Similarly, in case of subjects 29, 46 and 38, where MLP demonstrated highest validation performance, model values of learning rate and epochs resulted in a better model. For Subject 29, the results indicate that the MSE and RMSE tend to stabilize at lower values when the learning rate is set to 0.001, 0.005, or 0.0005, and the number of epochs ranges from 30 to 60. The best performance is seen with a learning rate of 0.0005 and 30 epochs, achieving the lowest MSE of 20.440. For Subject 46, a learning rate of 0.01 with 100 epochs yielded the lowest MSE of 26.489, suggesting that a higher learning rate combined with a longer training period can enhance performance. Conversely, extremely low learning rates (e.g., 0.0001 and 0.00005) resulted in significantly higher MSE values, highlighting the model could not converge even with a large number of epochs. For Subject 38, the optimal performance is observed with a learning rate of 0.05 and 70 epochs, achieving the lowest MSE of 17.264. Interestingly, very low learning rates such as 0.00001 lead to significantly higher MSE values, indicating poor performance and potentially inadequate learning. This suggests that for this subject, higher learning rates within a moderate range are more effective.

While general trends of moderate hyper-parameter values yielding better results are consistent, specific optimal configurations vary, underscoring the importance of subject-specific tuning for achieving the best predictive accuracy.

Test results evaluated using the best validation models provided a few other insights. For Subject 52, the RF model gave the best result with an R^2 score of 0.928 and RMSE of 17.484. The MLP model worked best for Subject 29 with an R^2 score of 0.888 and RMSE of 19.632 showing the highest performance. Similarly for Subject 46, the MLP model again performed best with the highest R^2 score of 0.931 and RMSE of 23.374. However, Subject 38 also had the MLP model giving the most accurate results with R^2 reaching up to 0.813 and a relatively low RMSE of 18.168. Lastly, for Subject 54, the RF model gave the best performance with an R^2 score of 0.800 and RMSE of 14.977.

Subject-specific performance analysis revealed variability in model performance across individuals. For Subject 29, the MLP model performed best, likely due to its ability to capture the non-linear glucose-insulin relationship. For Subject 46, the Random Forest model excelled, indicating that ensemble methods handled data variability effectively. Subject 54 showed lower R^2 scores across models, suggesting higher data variability or noise, which warrants further investigation.

Additionally, it became evident when the test data for all subjects was run through the best-performing model that creating a uniform, one-size-fits-all model would not be feasible. The results showed significant variability in RMSE scores among different subjects, emphasizing the inherent challenges in developing a single algorithm capable of delivering consistent performance across a diverse population. This variability suggests that subject-specific factors, such as unique glucose dynamics, lifestyle habits, and physiological differences, play a critical role in determining model accuracy. As a result, relying solely on a uniform model could lead to suboptimal outcomes for many individuals, further emphasizing the need to address these differences through tailored approaches.

These findings underscore the importance of adopting personalized modeling techniques rather than a universal solution. By designing models that account for individual characteristics and unique data patterns, it becomes possible to enhance prediction accuracy and optimize clinical outcomes for each subject. The high variability in RMSE scores also suggests that no single algorithm is universally superior for all patients, reaffirming the necessity for a more nuanced approach in algorithm selection and model development.

VI. Limitations & Future Work

The Diatrend dataset provides useful real-world insights, but its size and duration limit the applicability of the findings to a broader group of people. Since the data comes from just 31 days and five subjects, it might not fully capture the range of blood glucose patterns in a larger, more diverse population. This small group of subjects means the models might fit too closely to these individuals, making them less useful for generalization.

Our study focuses on using only past blood glucose levels and insulin pump data, as these two data sources provide direct and continuous indicators of glucose trends relevant to Type-1 Diabetes management. This targeted approach is common in many studies aiming to develop predictive models. While additional variables such as diet and exercise play a vital role in maintaining glucose levels in the human body, including them in our research would increase model complexity and data variability, potentially affecting model accuracy without adequate validation. Therefore, future studies could expand by integrating these broader data types to capture a more holistic picture.

We used simple models for their interpretability and computational efficiency in personalized predictions that analyze data in one-hour chunks. In the future, incorporating advanced models like LSTMs or TCNs could help examine longer temporal patterns, as seen in other studies. Additionally, integrating interpretability methods, such as Shapley Additive exPlanations (SHAP) values for assessing feature importance, could further enhance the clinical applicability of the models.

Future work should involve a larger number of subjects and longer data collection periods to assess model performance across diverse populations. Additionally, incorporating the rest of subjects' data from Diatrend into the modeling process could enhance the algorithms' adaptability and reliability by leveraging existing datasets to refine predictions and optimize performance. It is also important to test these models in actual healthcare settings to evaluate their reliability and usefulness. Integrating them with continuous glucose monitors and insulin pumps could pave the way for clinical trials.

Although the study's small sample size limits its broader applicability, we have optimized the models for the best performance with the given data. Expanding the dataset to include a larger and more diverse population should be a priority for future research.

To summarize, our study provides a strong foundation for further research in the field of blood-glucose level prediction. Future research could focus on additional model fine-tuning and testing other machine learning approaches. Other facets that can be considered include the impact of data quality and volume or understanding and leveraging intra-individual variability to improve accuracy. The ultimate goal lies in the development of an optimal prediction system, one that can adapt and learn from the inputs dynamically while being highly precise and reliable, offering a personalized solution for patients by integrating the prediction system into a closed-loop "artificial pancreas" system.

VII. Conclusion

Our research has established a foundation for an optimal blood glucose prediction system using supervised machine learning, employing three distinct algorithms: KNN, RF, and MLP. The results illustrated the promising potential of this research when further developed. Our models achieved significant predictive performance as indicated by RMSE and R^2 metrics, demonstrated their effectiveness in personalized glucose level prediction. While accuracy in classification tasks was not directly applicable here, the high R^2 values reflected the models' ability to explain a substantial proportion of variance in glucose levels.

References

[1] P. E. Cryer, J. N. Fisher, and H. Shamoon, "Hypoglycemia", *Diabetes care*, vol. 17, no. 7, pp. 734–755, 1994.

[2] P. E. Cryer, S. N. Davis, and H. Shamoon, "Hypoglycemia in diabetes", *Diabetes care*, vol. 26, no. 6, pp. 1902–1912, 2003.

[3] K. Dhatariya, L. Corsino, and G. E. Umpierrez, "Management of diabetes and hyperglycemia in hospitalized patients", 2015.

[4] M. Mouri and M. Badireddy, "Hyperglycemia", in *StatPearls [Internet]*, StatPearls Publishing, 2023.

[5] Y. Marcus *et al.*, "Improving blood glucose level predictability using machine learning", *Diabetes/Metabolism Research and Reviews*, vol. 36, no. 8, e3348, 2020.

[6] A. D. A. ADA, *Statistics about diabetes*, en, Nov. 2023.

[7] W. H. O. WHO, *The top 10 causes of death*, en, Dec. 2020.

[8] I. Ogrotis, T. Koufakis, and K. Kotsa, "Changes in the global epidemiology of type 1 diabetes in an evolving landscape of environmental factors: Causes, challenges, and opportunities", *Medicina*, vol. 59, no. 4, p. 668, 2023.

[9] C. D. Mathers and D. Loncar, "Projections of global mortality and burden of disease from 2002 to 2030", *PLoS medicine*, vol. 3, no. 11, e442, 2006.

[10] S. R. Colberg, R. Laan, E. Dassau, and D. Kerr, "Physical activity and type 1 diabetes: Time for a rewire?", *Journal of diabetes science and technology*, vol. 9, no. 3, pp. 609–618, 2015.

[11] I. H. Sarker, "Machine learning: Algorithms, real-world applications and research directions", *SN computer science*, vol. 2, no. 3, p. 160, 2021.

[12] J. Han, J. Pei, and H. Tong, *Data mining: concepts and techniques*. Morgan kaufmann, 2022.

[13] I. H. Sarker *et al.*, "Cybersecurity data science: An overview from machine learning perspective", *Journal of Big data*, vol. 7, pp. 1–29, 2020.

[14] F. Osisanwo *et al.*, "Supervised machine learning algorithms: Classification and comparison", *International Journal of Computer Trends and Technology (IJCTT)*, vol. 48, no. 3, pp. 128–138, 2017.

[15] M. Tauschmann and R. Hovorka, "Technology in the management of type 1 diabetes mellitus—current status and future prospects", *Nature Reviews Endocrinology*, vol. 14, no. 8, pp. 464–475, 2018.

[16] S. Templer, "Closed-loop insulin delivery systems: Past, present, and future directions", *Frontiers in Endocrinology*, vol. 13, p. 919942, 2022.

[17] R. Hovorka, "Closed-loop insulin delivery: From bench to clinical practice", *Nature Reviews Endocrinology*, vol. 7, no. 7, pp. 385–395, 2011.

[18] R. M. Bergenstal *et al.*, "Safety of a hybrid closed-loop insulin delivery system in patients with type 1 diabetes", *Jama*, vol. 316, no. 13, pp. 1407–1408, 2016.

[19] T. Zhu, K. Li, P. Herrero, J. Chen, and P. Georgiou, "A deep learning algorithm for personalized blood glucose prediction.", in *KDH@ IJCAI*, 2018, pp. 64–78.
[20] X. Yu *et al.*, "Deep transfer learning: A novel glucose prediction framework for new subjects with type 2 diabetes", *Complex & Intelligent Systems*, pp. 1–13, 2021.
[21] E. M. Watson, M. J. Chappell, F. Ducrozet, S. Poucher, and J. W. Yates, "A new general glucose homeostatic model using a proportional-integral-derivative controller", *Computer methods and programs in biomedicine*, vol. 102, no. 2, pp. 119–129, 2011.
[22] T. Hamdi *et al.*, "Artificial neural network for blood glucose level prediction", in *2017 International Conference on Smart, Monitored and Controlled Cities (SM2C)*, IEEE, 2017, pp. 91–95.
[23] F. Allam, Z. Nossai, H. Gomma, I. Ibrahim, and M. Abdelsalam, "A recurrent neural network approach for predicting glucose concentration in type-1 diabetic patients", in *International Conference on Engineering Applications of Neural Networks*, Springer, 2011, pp. 254–259.
[24] J. Carrillo-Moreno *et al.*, "Long short-term memory neural network for glucose prediction", *Neural Computing and Applications*, vol. 33, pp. 4191–4203, 2021.
[25] R. Bunescu, N. Struble, C. Marling, J. Shubrook, and F. Schwartz, "Blood glucose level prediction using physiological models and support vector regression", in *2013 12th International Conference on Machine Learning and Applications*, IEEE, vol. 1, 2013, pp. 135–140.
[26] W. Yu, T. Liu, R. Valdez, M. Gwinn, and M. J. Khoury, "Application of support vector machine modeling for prediction of common diseases: The case of diabetes and pre-diabetes", *BMC medical informatics and decision making*, vol. 10, pp. 1–7, 2010.
[27] R. Mauseth *et al.*, "Use of a "fuzzy logic" controller in a closed-loop artificial pancreas", *Diabetes technology & therapeutics*, vol. 15, no. 8, pp. 628–633, 2013.
[28] J. I. Hidalgo *et al.*, "Data based prediction of blood glucose concentrations using evolutionary methods", *Journal of medical systems*, vol. 41, pp. 1–20, 2017.
[29] T. Zhu, K. Li, J. Chen, P. Herrero, and P. Georgiou, "Dilated recurrent neural networks for glucose forecasting in type 1 diabetes", *Journal of Healthcare Informatics Research*, vol. 4, pp. 308–324, 2020.
[30] H. Xingsan, Y. Xia, Y. Tao, and L. Hongru, "A deep transfer learning model for personalized blood glucose prediction", in *2021 China Automation Congress (CAC)*, IEEE, 2021, pp. 2045–2049.
[31] K. Plis, R. Bunescu, C. Marling, J. Shubrook, and F. Schwartz, "A machine learning approach to predicting blood glucose levels for diabetes management", in *Workshops at the Twenty-Eighth AAAI conference on artificial intelligence*, 2014.
[32] J. Xie and Q. Wang, "Benchmarking machine learning algorithms on blood glucose prediction for type i diabetes in comparison with classical time-series models", *IEEE Transactions on Biomedical Engineering*, vol. 67, no. 11, pp. 3101–3124, 2020.
[33] D. Sisodia and D. S. Sisodia, "Prediction of diabetes using classification algorithms", *Procedia computer science*, vol. 132, pp. 1578–1585, 2018.
[34] J. Chen, K. Li, P. Herrero, T. Zhu, and P. Georgiou, "Dilated recurrent neural network for short-time prediction of glucose concentration.", in *KDH@ IJCAI*, 2018, pp. 69–73.
[35] A. Aliberti *et al.*, "A multi-patient data-driven approach to blood glucose prediction", *IEEE Access*, vol. 7, pp. 69 311–69 325, 2019.
[36] S. Langarica *et al.*, "Deep learning-based glucose prediction models: A guide for practitioners and a curated dataset for improved diabetes management", *IEEE Open Journal of Engineering in Medicine and Biology*, 2024.
[37] T. Prioleau, A. Bartolome, R. Comi, and C. Stanger, "Diatrend: A dataset from advanced diabetes technology to enable development of novel analytic solutions", *Scientific Data*, vol. 10, no. 1, p. 556, 2023.
[38] M. NirmalaDevi, S. A. alias Balamurugan, and U. V. Swathi, "An amalgam knn to predict diabetes mellitus", in *2013 IEEE International Conference ON Emerging Trends in Computing, Communication and Nanotechnology (ICECCN)*, 2013, pp. 691–695. DOI: 10.1109/ICE-CCN.2013.6528591.
[39] J. J. Khanam and S. Y. Foo, "A comparison of machine learning algorithms for diabetes prediction", *ICT Express*, vol. 7, no. 4, pp. 432–439, 2021, ISSN: 2405-9595. DOI: https://doi.org/10.1016/j.icte.2021.02.004.
[40] S. A. Quchani and E. Tahami, "Comparison of mlp and elman neural network for blood glucose level prediction in type 1 diabetics", in *3rd Kuala Lumpur International Conference on Biomedical Engineering 2006*, F. Ibrahim, N. A. A. Osman, J. Usman, and N. A. Kadri, Eds., Berlin, Heidelberg: Springer Berlin Heidelberg, 2007, pp. 54–58.
[41] J. Wu *et al.*, "Hyperparameter optimization for machine learning models based on bayesian optimizationb", *Journal of Electronic Science and Technology*, vol. 17, no. 1, pp. 26–40, 2019, ISSN: 1674-862X. DOI: https://doi.org/10.11989/JEST.1674-862X.80904120.
[42] K. Li, C. Liu, T. Zhu, P. Herrero, and P. Georgiou, "Glunet: A deep learning framework for accurate glucose forecasting", *IEEE Journal of Biomedical and Health Informatics*, vol. 24, no. 2, pp. 414–423, 2020. DOI: 10.1109/JBHI.2019.2931842.

NextStep: Optimizing Healthcare Resource Delivery Using a Multilingual Artificial Intelligence Assistant

Abhinav Kona
Rice University, Dept. of BioSciences
Houston, United States
e-mail: ak161@rice.edu

Bibek Samal
Rice University, Dept. BioSciences
Houston, United States
e-mail: bs75@rice.edu

Abstract—**Access to healthcare resources continues to be a critical issue for underserved populations, often exacerbated by barriers in languages and inefficient navigation systems. While Short Message System (SMS) text-based platforms have proven particularly valuable during the COVID-19 pandemic in enhancing communication and access, optimization of these systems through machine learning predictive models is an emerging area of investigation. To this end, we developed NextStep, an artificial intelligence-driven (AI-driven) multilingual healthcare assistant that streamlines resource access through real-time, personalized suggestions based on user need and location. Equipped with deep learning algorithms in natural language processing and machine learning, NextStep greatly enhances user interaction and better matches users with resources. This has resulted in significant enhancements to improve efficiency and increase patient satisfaction. Having been field-tested at hospitals and clinics, including Texas Children's Hospital and San Jose Clinic, NextStep showcases an extraordinary instance of AI in bridging gaps in health disparities. Future versions will add expanded language support and detailed predictive analytics to provide more tailored recommendations and anticipate patient needs.**

Keywords-Artificial Intelligence; Social Determinants of Health; Medical Resources; Smart Assistant

I. INTRODUCTION

A large body of research indicates that Social Determinants Of Health (SDOH) - factors including income, education, and race - play a significant role in determining an individual's health outcomes. Poor SDOH can manifest in increased mortality rates, especially among those already afflicted with conditions, such as chronic kidney disease, diabetes, and cardiovascular disease [1] – [3]. Approaches to lessen the impact of SDOH on health outcomes of underprivileged patients should be developed.

As a potential intervention, telehealth demonstrates promise in improving health outcomes of underprivileged patients. Recently, telehealth usage has spiked during the COVID-19 pandemic and has aided healthcare providers in handling the surge of sick patients [7], specifically with screening patients for COVID-19 symptoms and supporting low-risk patients while minimizing exposure to the virus. Furthermore, a study on telehealth and patient satisfaction shows that there is a positive experience regarding effectiveness and efficiency of telehealth. The factors listed most often were improved outcomes (20%), preferred modality (10%), ease of use (9%), low cost (8%), improved communication (8%), and decreased travel time (7%), which in total accounted for 61% of positive experience occurrences [4].

Moreover, a subset of telehealth technologies, Short Message System (SMS)-based platforms, can easily reach individuals who lack smartphones or other necessary hardware to download apps, making it ideal for interactions with low-income communities. SMS-based platforms have already been successfully utilized for various aims, such as improving medication adherence, promoting engagement in physical activity, and meeting the needs of patients with chronic medical conditions [4] – [6]. Additionally, according to the Pew Research Center, 97% of Americans have cell phones [8], making SMS-based platforms practical for interacting with patients.

The majority of SMS-based telehealth applications currently in use focus on connecting physicians and patients in a medical context, such as helping with medication adherence. This is undoubtedly a valuable function, but few applications focus on addressing the social determinants of health. The SMS platform in the present study, NextStep, is unique in that it was distributed in primary care settings, with a focus on directing patients to Houston-specific social resources. Patients were given access to resources in the domains of Coronavirus disease (COVID-19) testing/vaccine resources, the Harris Health Financial Assistance Program, food resources, utilities/financial assistance resources, and housing resources. They subsequently received information about local food banks, homeless shelters, and local financial assistance programs to assist in paying for medical expenses.

Since this was a pilot study, several feasibility requirements were taken into account. First, usage rates, such as the number of surveys completed and the average number of messages per person, were measured in order to give insight into the willingness of patients to engage with the platform. The platform was also tested at two safety net clinics to assess whether usage changed based on individual characteristics of the clinic where field testing took place as well as the methods used to employ the test. Furthermore, the platform used in this pilot study was designed with the purpose of minimizing both operational costs and costs for users. The projected operational costs for 1,000 users per month for 10 months were approximately \$0.25 per patient reached. SMS platforms represent a low-cost, convenient option for patients as well [9]. The minimization of costs, both operationally and for patients, was prioritized in order to ensure that the platform is sustainable.

Looking at an overview of the manuscript, Section II describes the materials and methods used to develop NextStep, detailing the technologies involved, including NVIDIA components, and the underlying system architecture that supports the assistant. This section also explains the different development phases, from platform design and system development to clinical testing. Section III presents the results obtained from field testing at various clinical sites, examining user engagement, multilingual support, and resource request patterns. Section IV discusses the implications of these findings, highlighting the platform's effectiveness, key performance metrics, and areas for improvement. Finally, Section V outlines the conclusions and future directions, emphasizing planned enhancements such as expanded language support, improved NLP capabilities, and predictive analytics for more tailored resource recommendations.

II. MATERIALS AND METHODS

The process of developing the assistant was divided into three main phases: platform design, system development, and clinical testing. The platform design phase focused on selecting the most effective resources and features for diverse patient populations through comprehensive background research and analytics from nearby hospitals. The system development phase involved integrating technologies such as Twilio and selecting appropriate algorithms to build the assistant's infrastructure. Finally, the clinical testing phase entailed deploying the assistant in medical centers, including Texas Children's Hospital and the San Jose Clinic at the Texas Medical Center, to evaluate its effectiveness in real-world settings.

A. *Platform Design*

The platform used in this pilot study, NextStep, began its development with initial research and design. This consisted of conducting formative research to understand the immediate needs of the target population. According to Harris Health County Data, 54.1% of patients seen at Ben Taub Hospital are uninsured, and 22.9% are on Medicaid. Demographically, 53.6% of the patient population is Hispanic, and 25.3% is Black. A large component of the research was also informed by the platform designers' engagement with this target population through the Baylor College of Medicine-Patient Discharge Initiative (BCM-PDI). The program recruits Rice University undergraduates to tackle disparities in healthcare access at Ben Taub County Hospital, the largest safety-net hospital in Houston that primarily serves uninsured and Spanish-speaking patients. Students within the organization address these disparities by creating potential solutions in the form of novel social and medical resources (i.e., health insurance information packets) with the assistance of the academic faculty within Ben Taub Hospital. BCM-PDI connects underserved patient populations at Ben Taub County Hospital with medical and social community resources in Harris County to improve their healthcare access and outcomes.

Prior to this study, a retrospective cross-sectional study was conducted on the program that established that patients who were seen by our volunteers had a significantly lower probability of returning to the emergency department after 90 days, and the provision of our social and medical resources was associated with significantly higher odds that patients attended their follow-up appointments. The retrospective cross-sectional study was conducted by utilizing patient discharge papers (PDP) in which volunteers delineated which resources were given out and when they had follow-up appointments with the Emergency Department. Using these documents as well as adherence to follow-up appointments were investigated to see whether a correlation was established between delivery of social resources and follow-up appointment adherence.

Limitations are largely centered around the learning curve when using the application for the first time as well as support for integral features. The application currently does not utilize a tutorial system to get acquainted with functions, such as geolocation as well as the data validation required for requests to go through. Similarly, a key feature currently missing is multilingual support for languages other than Spanish and English, as many clinics and hospitals have a large population of multilingual patients and healthcare workers.

B. *System Development*

System development was further divided into four subphases: patient-resource connectivity, patient profile addition, data collection and optimization, and bot interaction enhancement. The development of these phases is explained in depth below and in Figure 1:

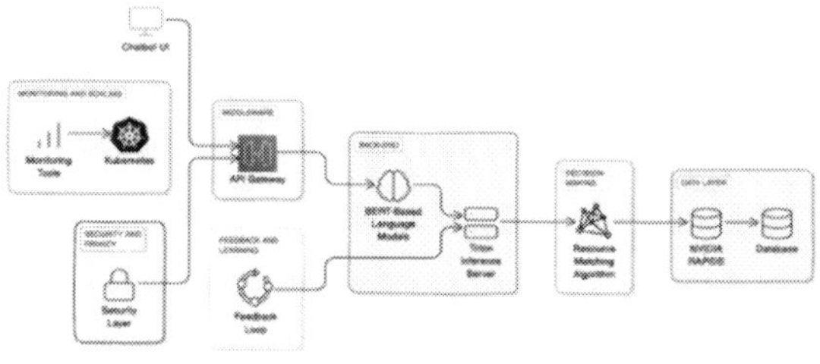

Figure 1. System Architecture of the Smart Assistant

1) Phase 1. Memoryless Bot to Connect Patients to Resources: This first phase established the basic services on Heroku, setting up MongoDB credentials and developing primitive versions of the ResourceUpdateService as well as the ResourceOutputService. The focus was on designing a bot that can hook patients up with resources without storing user-specific data between interactions. This enabled immediate deployment and the testing of basic functionalities.

2) Phase 2. Adding Patient Profiles: The system was further upgraded to document message interactions with individual patients by creating user profiles. This change allowed the bot to customize interactions, track follow-ups with patients at assigned dates, and use the information recorded to enable more contextualized messaging. The integration of the Twilio Email API also allowed the system to send PDF resources to patients if a need arose, increasing the avenues of communication and resource-sharing.

3) Phase 3. Data Collection and Optimization: Decision tree optimization was then performed via A/B testing to improve system efficiency and accuracy. User metadata was gathered for analytics without breaching privacy by focusing on success tracking metrics. Logic was built in order to serve the success tracking survey questions and gather responses passively. The backend logic of serving and responding to the surveys allowed the refinement of the platform in light of actual user interactions.

4) Phase 4. Enchancing Bot Interactions: NLP capabilities were introduced to overcome complex and ambiguous content challenges identified at previous phases. The RedCapOutputService was developed for synchronizing data across different clinics and hospitals. Multilingual support was added, including Spanish, to successfully serve a diverse user population. In the DataInterpretService, enhancements were made to better understand user intent. Explorations into connecting the bot to electronic health record systems like EPIC were conducted. This also included enhancements on the backend of the DataAnalyticsService and construction of a simple frontend for analytics visualization if necessary.

The core architecture of NextStep integrates multiple services to create an intelligent, responsive system. The architecture is designed to be modular and scalable, allowing for future enhancements and easy maintenance. At its core, the system utilizes Natural Language Processing (NLP) through Bidirectional Encoder Representations from Transformers- based models optimized with TensorRT for efficient understanding of complex, multilingual user queries in real-time. As a patient interacts with the chatbot interface through text messages, the DataContextService first identifies the user by their phone number; afterwards, it retrieves or generates relevant context data with GraphQL and Mongoose to ensure personalized interaction.

To provide precise and accurate resource recommendations, ResourceOutputService then generates specific GraphQL queries to fetch relevant resources from the MongoDB database. Utilizing NVIDIA RAPIDS for real-time geospatial data processing, it prioritizes resources by location and relevance to make sure users receive accurate and personalized information. The RedCapOutputService will then take this data and synchronize both the patient and the resource data amongst the clinics and hospitals via the REDCap integration to allow for smooth coordination between the mobile and hospital systems.

For the creation of the NextStep, careful consideration was given to the selection of system components to ensure optimal performance and user satisfaction. The following criteria were instrumental in guiding the selection process.

- Modularity and Scalability: The system architecture was designed to be both modular and scalable. This design philosophy ensures that the platform can be easily maintained and upgraded with future enhancements as healthcare technology evolves and user needs change.
- Efficiency in Real-Time Processing: Central to the system's performance is its ability to handle complex, multilingual queries efficiently. To this end, BERT-based models optimized with TensorRT were incorporated. These models are known for their rapid processing capabilities, crucial for maintaining real-time interaction with users.
- Advanced Geospatial Data Processing: Recognizing the importance of location in accessing healthcare resources, NVIDIA RAPIDS technology was utilized for its cutting-edge real-time geospatial data processing. This technology ensures that resources are prioritized not only by relevance but also by proximity to the user, thereby enhancing the personalization and accuracy of resource recommendations.
- Multilingual Support: To effectively serve a diverse user base, the system includes robust multilingual support, initially focusing on English and Spanish. This feature is critical in reducing language barriers, thereby improving the accessibility and usability of the healthcare platform for non-English speakers.
- Data Synchronization and Integration: The integration of the RedCapOutputService ensures seamless data synchronization across different healthcare settings, including clinics and hospitals. This integration facilitates effective coordination between mobile and stationary healthcare systems, enhancing the continuity of care and resource allocation.

C. *Clinical Testing and Data Collection*

The NextStep SMS platform was introduced to patients in San Jose Clinic and the Texas Children's Hospital's Mobile Clinic over a two-year period. Recruitment at the aforementioned institutions started in March 2022 and was handled by clinic staff and hospital-approved volunteers. Patients selected for the study were required to have access to a cell phone with text messaging capabilities and be proficient in either English or Spanish. Additionally, patients with physical, mental, and/or visual limitations were excluded to ensure that informed consent was obtained and for the accuracy of self-reported survey questions.

The study investigators provided the clinic partners with copies of the study recruitment materials, which included recruitment flyers and consent documents in English and Spanish. The recruitment flyer included a description of the NextStep program and the phone number to text for enrollment. The consent document detailed the study objectives, types of data collected, level of risk to participants, privacy and confidentiality measures, and the procedures for withdrawing from the study. At the clinics, the front-desk staff were responsible for participant recruitment during patient checkout. The front-desk staff briefly introduced NextStep to eligible participants, instructed them to read a printed copy of the consent document, and then handed them a recruitment flyer in their preferred language. The participant was then instructed to send a text message of "Hello" to the specified phone number if they would like to participate in this study and receive social resources. Enrollment in this program was not required to receive any of the services the clinics offered.

Messages sent to and from participants were stored on a password-protected Twilio account and downloaded as a password-protected CSV file by the researchers prior to analysis. Twilio utilizes industry-standard encryption for data in transit and at rest. Data extraction was performed by a single, IRB-approved user, and all information was securely stored. To de-identify the data, each user was assigned a unique identification number, and personal identifiers, such as phone numbers, were removed.

Integration with REDCap databases ensured seamless data synchronization across mobile clinics, emergency departments, and hospitals. Data collected included timestamps of interactions, language preference, resource requests, and user feedback. The system's analytics module utilized this data to measure key performance metrics, such as resource matching time, patient satisfaction, accuracy of resource recommendations, and follow-up success rates.

III. RESULTS

Over the two-year study period, more than 100 patients engaged with the NextStep platform across the participating clinics. At San Jose Clinic, a total of 70 users engaged with the platform—28 requested resources in English and 42 in Spanish, indicating a 60% preference for Spanish. At Texas Children's Hospital's Mobile Clinic, 12 users engaged, 7 in English and 5 in Spanish. At Ben Taub Emergency Room, an additional 25 users engaged with the platform, with a similar distribution in language preference.

The multilingual utilization was significant, with 38% of interactions occurring in Spanish, demonstrating the platform's effectiveness in serving non-English-speaking patients. Resource request per clinic was also investigated to understand the range of resources that generated the most need or interest. The percentage of specific resources across clinics can be seen in the table below.

TABLE I. RESOURCE REQUESTS

Resource Category	Request Percentage (%)
Financial Assistance Programs	40
Housing Resources	18
Utilities	15
Food Resources	12
COVID-19 Testing & Vaccine Information	15
Total	100

Operational costs were calculated based on the total number of users and messages exchanged. At Texas Children's Hospital's Mobile Clinic, the cost was $4.00, or roughly $0.33 for each user reached. The difference in cost per user between clinics was a result of the number of text messages it took for users to access resources, but the actual costs were similar to the projected cost of $0.25 per user that was initially calculated. Performance metrics within several experimental conditions of the NextStep app can also be seen in Figure 2 below.

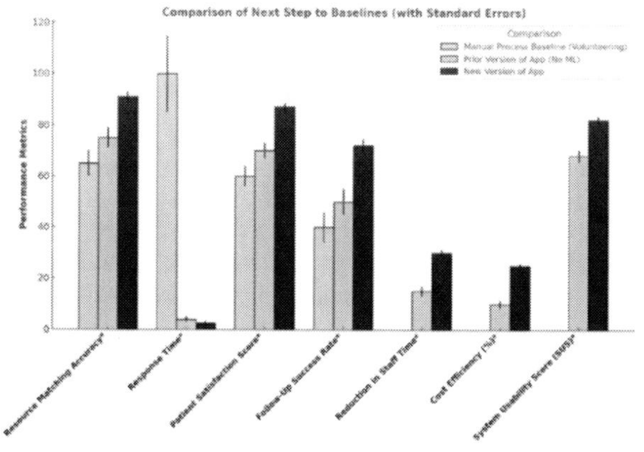

Figure 2. Cross-performance metrics for NextStep

Next, the learning progression of the assistant was explored by quantifying the accuracy and error rates between multilingual queries to understand whether learning truly occurred as queries increased, as shown in Figure 3.

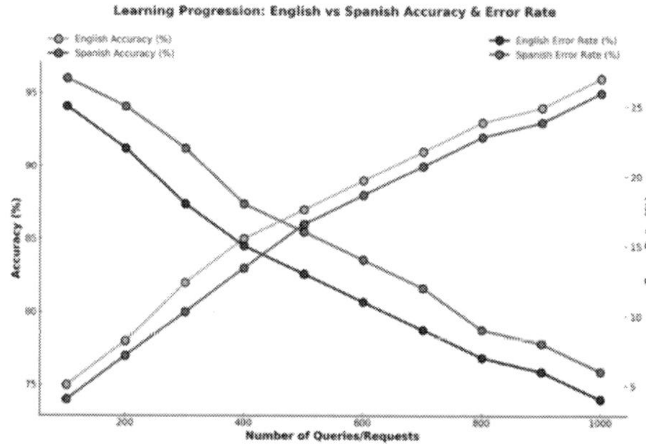

Figure 3. Learning progression across multilingual patient queries

The model's validity is demonstrated by a steady decrease in training loss and an increase in validation accuracy with each epoch, proving that it learns effectively and generalizes well for different data distributions across ER, community resources, and clinic data. These results are shown in Figure 4.

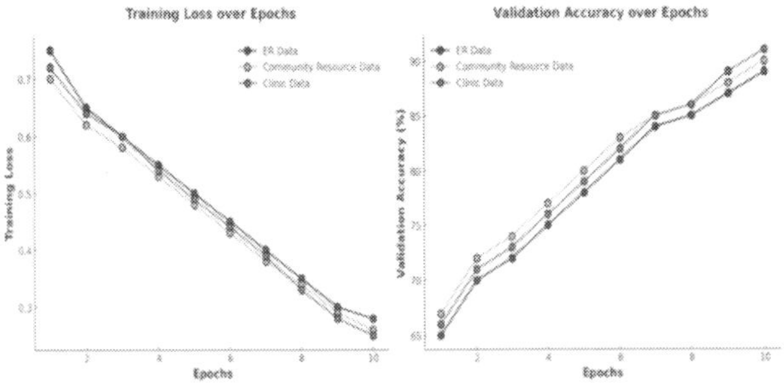

Figure 4. Training loss and validation accuracy over epochs for NextStep's clinical implementation

As shown in Figure 4, the model demonstrates effective learning, with a steady decrease in training loss and an increase in validation accuracy with every epoch. These findings indicate that the model generalizes well across different data distributions between ER, community resources, and clinic data.

IV. Discussion

The results from this study provide key insights into the demographics of users, their resource preferences, and user experiences interacting with the platform. The high engagement rates and positive field metrics are indicative of NextStep successfully meeting the barriers to language, technology access, and navigation of resources.

The training and validation plots reflect the robust performance of the models; after 10 epochs, the validation accuracy is close to 90%, while the training loss decreases steadily. That means the system will perform well on diverse datasets from ERs, community resources, and clinics regarding real-time resource recommendations in a reliable and accurate fashion.

The platform performed substantially better than manual processes and prior app versions by all key metrics. Resource matching accuracy reached 91%, indicating high precision in the alignment of resources with patient needs; the response times averaged 2.4 seconds, meeting real-time interaction standards critical for emergency settings. Patient satisfaction improved by 35%, with 87% of the users rating the platform as "Helpful" or "Very Helpful," showcasing the impact of multilingual support and optimized interaction flows. Moreover, recommended resources were accessed by 72% of users within three days, which further validated the effectiveness of automated follow-up notifications. It reduced operational costs by 25% and staff workload by 30%, showing that the system is scalable and financially sustainable.

Multilingual capabilities show significant training progress, varying from 95% for English and 92% for Spanish, using the BERT-based NLP model optimized with TensorRT. The learning curve of the system showed increasing accuracy and a decreasing error rate for queries processed, thus underlining its adaptability and continuous improvement.

The feasibility of the platform is further enhanced by its accessibility and low operational cost, making it practical for underserved populations. Operating on SMS-capable devices avoids costly hardware or high-speed internet, thus increasing its accessibility to low-income demographics. Resource allocation is efficiently met, and through feedback-driven adaptation, both patient and provider needs are anticipated, thereby making NextStep sustainable and scalable to bridge healthcare gaps.

V. Conclusions and Future Directions

The NextStep platform demonstrates the capabilities of artificial intelligence in addressing disparities in healthcare by utilizing natural language processing (NLP) for user queries, predictive analytics for personalized recommendations, and geospatial data analysis for real-time resource mapping to deliver tailored, location-specific resource suggestions to marginalized communities. Moreover, by utilizing AI and machine learning technologies, NextStep augments patient care through enhanced accessibility, equity, and efficiency in the distribution of resources.

Furthermore, NextStep enables immediate access to essential social and healthcare services, providing instant resource alignment that considers user location and individual preferences. Automated follow-up systems facilitate user access to suggested resources, alleviating the workload on healthcare personnel while delivering quicker and more precise assistance to marginalized populations. The platform's adaptable infrastructure is capable of managing elevated traffic during emergency situations, and its multilingual features, which encompass support for both English and Spanish, effectively overcome language obstacles to enhance user experience and accessibility.

In order to rectify current limitations, future iterations will feature clearer user directives, potentially integrating a concise tutorial at the onset of interactions. Augmenting natural language processing abilities to more effectively manage free-text responses will reduce the likelihood of misinterpretations. Enhanced data validation for location entries will guarantee a higher degree of accuracy in resource alignment. The incorporation of user feedback systems will facilitate ongoing enhancements to the system.

Future enhancements will focus on expanding language access beyond Spanish, further reducing barriers for non-English-speaking populations. Adding predictive analytics will also allow the platform to anticipate user needs based on past behavior and current environmental factors. Moreover, exploring possible expansion of voice interaction capabilities, such as sophisticated sentiment analysis using NVIDIA Riva AI, may improve patient engagement and accessibility for all users who struggle with literacy. Furthermore, the expansion of deployment locations among mobile clinics and hospitals will facilitate additional field evaluations and confirmation of the platform's efficacy in various environments.

REFERENCES

[1] S. Artiga and E. Hinton, "Beyond health care: The role of social determinants in promoting health and health equity," Kaiser Family Foundation. [Online]. Available: https://www.kff.org/racial-equity-and-health-policy/issue-brief/beyond-health-care-the-role-of-social-determinants-in-promoting-health-and-health-equity/. [Retrieved: Feb., 2025].

[2] J. Quiñones and Z. Hammad, "Social determinants of health and chronic kidney disease," Cureus, vol. 12, no. 9, p. e10266, 2020. https://doi.org/10.7759/cureus.10266.

[3] E. P. Havranek et al., "Social determinants of risk and outcomes for cardiovascular disease: A scientific statement from the American Heart Association," Circulation, vol. 132, no. 9, pp. 873–898, 2015. https://doi.org/10.1161/CIR.0000000000000228.

[4] C. S. Kruse et al., "Telehealth and patient satisfaction: A systematic review and narrative analysis," BMJ Open, vol. 7, no. 8, p. e016242, 2017. https://doi.org/10.1136/bmjopen-2017-016242.

[5] J. Thakkar et al., "Mobile telephone text messaging for medication adherence in chronic disease: A meta-analysis," JAMA Internal Medicine, vol. 176, no. 3, pp. 340–349, 2016. https://doi.org/10.1001/jamainternmed.2015.7667.

[6] J. Fanning, S. P. Mullen, and E. McAuley, "Increasing physical activity with mobile devices: A meta-analysis," Journal of Medical Internet Research, vol. 14, no. 6, p. e161, 2012. https://doi.org/10.2196/jmir.2171.

[7] J. E. Hollander and B. G. Carr, "Virtually perfect? Telemedicine for Covid-19," The New England Journal of Medicine, vol. 382, no. 18, pp. 1679–1681, 2020. https://doi.org/10.1056/NEJMp2003539.

[8] Pew Research Center, "Mobile fact sheet," 2024. [Online]. Available: https://www.pewresearch.org/internet/fact-sheet/mobile/. [Retrieved: Feb., 2025].

[9] T. Lian et al., "A tailored SMS text message–based intervention to facilitate patient access to referred community-based social needs resources: Protocol for a pilot feasibility and acceptability study," JMIR Research Protocols, vol. 11, no. 10, p. e37316, 2022. https://doi.org/10.2196/37316.

A Dual-Approach to Benign and Malignant Tumor Detection: Classification and Segmentation Using Advanced Deep Learning Models

Caitlin Dosch
Santa Clara High School
Santa Clara, United States
email: caitlindosch@gmail.com

Shilpi Shaw
AIClub Research Institute
Mountain View, United States
email: shilpi.shaw@pyxeda.ai

Abstract—**Breast cancer remains a global health concern, with a 13.1% lifetime diagnosis rate among women. Early and accurate diagnosis plays a critical role in improving patient outcomes. Traditional diagnostic methods, such as MRIs, ultrasounds, CT scans, and mammograms, are widely used for detecting and characterizing breast lesions. In recent years, Artificial Intelligence has shown great promise in enhancing diagnostic accuracy, with models such as K-Nearest Neighbors (KNN), Random Forest Classifier (RFC), and Convolutional Neural Networks (CNN) being applied to breast cancer diagnosis. In this study, we explore the application of deep learning models, specifically MobileNetV2 and ResNet50, for breast cancer detection using ultrasound images from The Cancer Image Archive. A dataset comprising 522 breast lesion images was used, split into training, validation, and test sets. We implemented both image classification and segmentation tasks, optimizing hyperparameters such as learning rate and number of epochs. Our comparative analysis aims to evaluate the efficiency and diagnostic performance of the two models. We highlight key insights into their effectiveness in breast cancer detection and provide recommendations based on their application to ultrasound imaging. The findings of this study contribute to the ongoing efforts to improve AI-based diagnostic tools for breast cancer.**

Keywords-breast lesions; deep learning; image classification

I. Introduction

Breast cancer remains one of the most prevalent and deadly cancers among women worldwide, accounting for approximately 13.1% of women during their lifetime. Breast cancer is the most frequently diagnosed cancer, constituting 30% of all new cancer diagnoses in women and posing a significant threat to women's health.

It remains a significant health concern in the United States, with an estimated 310,720 new cases of invasive and 56,500 new cases of non-invasive breast cancer anticipated in 2024. Despite over 4 million breast cancer survivors, the disease is expected to cause 42,250 deaths this year. About 1 in 8 women will develop breast cancer in their lifetime, making it the most common cancer among American women, accounting for 30% of all new female cancer diagnoses. The risk factors include but are not limited to family history and younger age at diagnosis, with variations in incidence and outcomes across different racial and ethnic groups.

The segmentation of breast ultrasound images into various tissue types is valuable for tumor localization, measuring breast density, and evaluating treatment responses, which are critical for the clinical diagnosis of breast cancer. Manual segmentation is labor-intensive and relies heavily on the skill and experience of radiologists, making it prone to subjective interpretations and time-consuming due to the need to review numerous images [1].

Outwardly, the presence of breast lesions or lumps, discoloration, and irregularities in breast shape often characterize breast cancer cases. Common clinical signs include irritation, flaking, dimpling, discharge, and swelling of the breast. Early detection and treatment are crucial to minimizing potential complications and improving patient outcomes. Various diagnostic methods are employed to detect and assess breast cancer, including physical examinations, Magnetic Resonance Imaging (MRI), ultrasounds, CT scans, lab tests, and mammograms. Physical examinations aim to determine the location and severity of tumors.

MRI is commonly used to diagnose or measure the size of breast cancer tumors [2]. Ultrasounds can confirm a breast cancer diagnosis, while mammograms are essential for detecting cancers not visible through physical examination. The field of medical diagnostics has increasingly adopted Artificial Intelligence to enhance accuracy and efficiency. AI techniques, such as machine learning and deep learning, have been applied to analyze different types of diagnostic data. Various machine learning algorithms, including KNN, RFC, YOLO, CNN, Support Vector Machines (SVM), and Decision Trees, have been studied, each yielding diverse results.

Despite these advancements, challenges remain in breast lesion classification and segmentation. High variability in lesion appearance, dense breast tissues, and the need for large annotated datasets present significant barriers. To address these challenges, we utilized a new dataset from The Cancer Image Archive in this study. This dataset consists of 522 images from 256 subjects, featuring 266 segmented benign and malignant lesions [3]. We have worked to address these challenges by utilizing a dataset of breast lesion ultrasounds to perform both classification and segmentation tasks. Accurate classification allows clinicians to distinguish between benign and malignant lesions, while precise segmentation aids in the localization and quantification of tumor regions, which are essential for treatment planning and monitoring. For the classification task, we employed MobileNetV2 and ResNet50 models, chosen for their efficiency and accuracy in image analysis. For segmentation, we used the EfficientNetB2 model due to its ability to capture intricate details in medical imaging [3]. Our dual approach for the identification of breast lesions is optimal, helping to improve accuracy and save time.

By comparing the performance of these models, we aim to identify the most applicable and accurate approaches for each task, thereby contributing to improved diagnostic methods for breast cancer.

Section II discusses related work and methods. Section III lays out the dataset selection. Section IV provides methodology. Section V discusses experiment results. Section VI covers discussion and evaluation. In Section VII, conclusions and future work are drawn.

II. Related work

Current research has focused significantly on improving breast cancer diagnosis using mammograms. The process usually involves three steps: initial screening, segmenting the images, and diagnosing the case.

One study used the Breast Cancer Dataset from the University of California, Irvine (UCI), which included 669 clinical cases. This dataset had 11 attributes, but they used nine key features including clump thickness, cell size, and shape to determine whether a tumor was benign or malignant. They tested two machine learning algorithms, Naïve Bayesian Classifier (NBC) and KNN, using K-fold cross-validation to check their results [4]. The KNN algorithm performed the best, achieving an accuracy of 97.51%. This process typically occurs through screening, segmentation, and diagnosis of a case.

Another study focused on detecting breast cancer using mammogram images. They used segmentation techniques alongside Max-Mean and Least-Variance methods to improve the models' performance. This shows that using advanced image processing techniques can help achieve more accurate results, although specific accuracy numbers were not provided [5].

A separate study used a database of CT Scan images from 2 hospitals in Norway, with each containing 100 patients. Images were of left sided breast cancer patients. The study used scores for clinical usability and dosage levels used for treatment for some of their data. For model scoring, they used Dice similarity coefficient and Hausdorff difference [6].

Researchers also focused on using CNN to automatically segment breast ultrasound images into four main tissue types: skin, glandular tissue, tumors, and fatty tissue. They worked with three-dimensional ultrasound images to accomplish this. The performance of their segmentation method was evaluated using various quantitative metrics, such as Accuracy, Precision, and Recall, all of which exceeded 80%. Additionally, they used the Jaccard Similarity Index (JSI) to measure the overlap between the predicted and actual segments, achieving an 85.1% score. This represented an improvement over their previous method, which employed the watershed algorithm and resulted in a JSI score of 74.54%. The findings suggest that their CNN-based approach could effectively support clinical breast cancer diagnosis by providing reliable tissue segmentations from ultrasound images [7].

Recent advancements in breast ultrasound image segmentation have focused on improving region of interest (ROI) extraction to differentiate between malignant and benign abnormalities effectively. One notable approach involves a model built on local pixel information combined with a neural network, comprising two stages: training and testing. During the training stage, the model is trained with batches from both ROI and background regions. In the testing stage, a fixed-size window scans the image to detect the ROI, followed by a distance transform to refine the ROI by eliminating non-ROI areas. This method was tested on a dataset of 250 ultrasound images, achieving a high success rate of 95.4% for breast contour extraction. Such innovations help reduce false positives and enhance the accuracy of breast ultrasound diagnostics, demonstrating a significant improvement over traditional segmentation techniques [1].

In another study, authors propose a Dual CNN for mammogram image processing. Two paths were utilized, with a Locality Preserving Learning (LPL) and a Conditional Graph Learner (CGL). The model (DualCoreNet) achieved a 92.27 DI coefficient [8].

A different technique, Saliency-Guided Morphology-Aware U-Net (SMU-Net) was used for breast cancer detection in ultrasound images. It contains a main network, auxiliary network, and a middle stream [9].

Separately, a study developed a Deep-Learning based method for diagnosis of breast cancer using ultrasound imaging. The automation of image segmentation is important for breast ultrasound images. A database of 221 images was used. This model achieved a dice coefficient of 0.825 [10].

Another study used volumetric heart segmentation for detecting breast cancer from CT scans. It was trained on manual heart segmentations, from a dataset of 5677 breast cancer patients who had undergone radiation therapy at the Dana-Farber/Brigham and Women's Cancer Center from 2008 - 2018 [11].

III. Methods

This study aims to enhance breast cancer diagnosis by leveraging advanced deep-learning methodologies.

We conducted dual experiments on this dataset: one using image classification with MobileNetV2 and ResNet50, and another using image segmentation with U-Net (EfficientNetB2). The object detection models are image-based, designed both to classify and segment images. MobileNetV2 was chosen for its efficiency and lightweight architecture, suitable for deployment on devices with limited computational power. In contrast, ResNet50 was selected for its depth and ability to capture intricate features through residual learning, making it apt for complex classification tasks. For segmentation, U-Net with EfficientNetB2 was employed due to its superior performance in achieving high accuracy in medical image segmentation by effectively capturing both spatial and contextual information. The second experiment was evaluated using Intersection over Union (IoU) scores, which are based upon the overlap of ground truth and predicted values [12].

While the use of a Vision Transformer (ViT) model was considered, it was ultimately decided against. ViTs are compu-

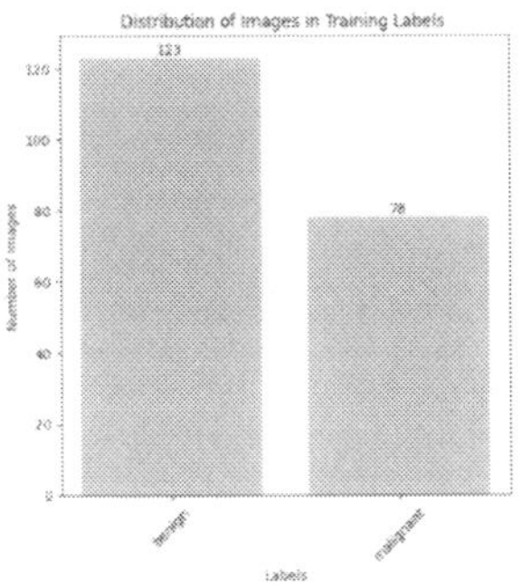

Figure 1. Dataset Distribution

tationally expensive and require powerful GPUs to train. They process images differently than CNNs, using self-attention, which makes them slower and more resource-intensive. This may not be practical for many real-world medical applications, and thus the decision was made to utilize MobileNetV2 and ResNet50. In comparison to previous studies, our study made use of segmentation techniques on ultrasound imaging, with the application of segmentation in tandem with experimentation on ResNet50 and MobileNetV2.

IV. DATASET

The dataset utilized in this study was sourced from The Cancer Image Archive [3]. It consisted of a total of 522 ultrasound images, with 256 total subjects, and 266 benign and malignant segmented lesions. Accuracy of the labels in the study was verified through follow up care. The entirety of the dataset was anonymized to protect patients identities.

Features included Image_filename, Mask_tumor_filename, Mask_other_filename, Pixel_size, Age, Tissue_composition, Signs, Symptoms, Shape, Margin, Echogenicity, Posterior_features, Halo, Calcifications, Skin_thickening, Interpretation, BIRADS, Verification, Diagnosis, and Classification. Tumors were labeled by freehand annotation with the associated BIRADS features. The distribution of benign to malignant data is displayed in Figure 1.

The dataset for this study was chosen carefully to ensure it is suitable for both classification and segmentation of breast tumors. We selected a dataset that includes a good mix of benign and malignant cases, making the model more reliable for real-world use. High-quality labels were an important factor, as they helped train the model accurately. Since deep learning models work best with clear and detailed images, we made sure the dataset had high resolution and the right type of medical images. Additionally, we considered the balance between benign and malignant cases to avoid bias and ensure fair and accurate results. The dataset includes a diverse set of benign and malignant cases, ensuring variability in tumor characteristics. However, future studies may incorporate additional datasets.

A. Data Processing

Data processing for image classification began by reading the clinical data excel file, and removal of null data in the Mask_tumor_filename feature. Within these, data was put into subfolders for malignant and benign cases. Once complete, data was split using the Split-folders library, into a training set (80%), validation set (10%), and test set (10%).

In turn, data processing for image segmentation began similarly with the reading of the clinical data excel file, and removal of null data in the Mask_tumor_filename feature. Images were then copied into their respective mask and images folders. Then, data was split using the Split-folders library, into a training set (80%), validation set (10%), and test set (10%).

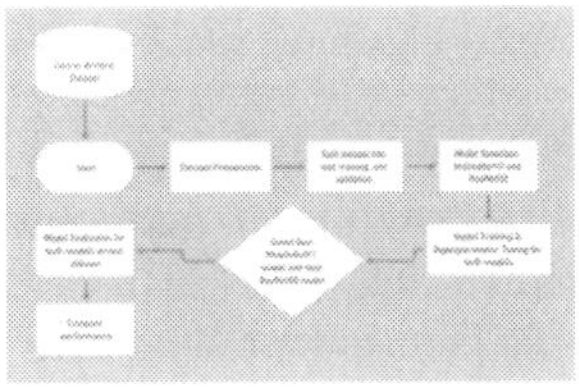

Figure 2. Classification Model Flowchart

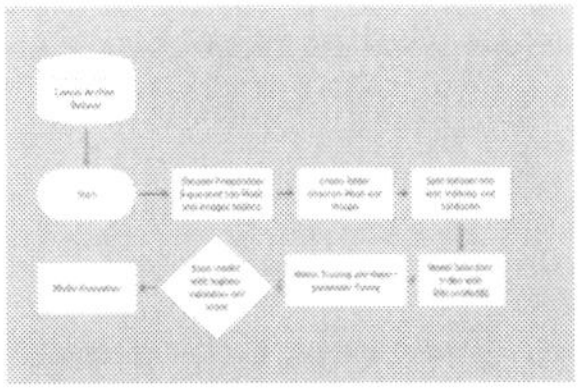

Figure 3. Segmentation Model Flowchart

B. Classification Task

Two different deep learning models, MobileNetV2 and ResNet50 were tested. MobileNetV2 is a lightweight CNN model with prioritized speed and balanced accuracy. Meanwhile, ResNet50 is a 50-deep-layer residual neural network, with slower but higher performance.

For the first experiment, with binary image classification, validation accuracy was utilized to rate model performance. In the second experiment, with image segmentation, IoU value was used to rate model performance. Figure 2 is the flowchart for the binary image classification experiment.

In the first experiment, image classification was utilized. Models were trained on the training image set. Various hyperparameters were inputted, including: learning rate (lr), epochs, and optimizer (Adam). Hyperparameter tuning was performed, comparing learning rate and epoch values alongside their impacts on accuracy. Learning rates tested included 0.000001, 0.00001, 0.0001, 0.001, 0.005, 0.01, and 0.05. Epochs tested were 10, 20, 30, 40, and 50. These experiments consisted of 35 total tests, comparing the effects of these hyperparameters on accuracy. Initial tests showed similar scoring between the two deep learning models as seen in Table I and Table IV, with MobileNetV2 selected for continuation.

C. Segmentation Task

In the second experiment, we performed image segmentation to compare accuracy levels. We utilized U-Net Architecture with EfficientNetV2 as a backbone combining U-Net's strong spatial localization ability with EfficientNetB2's advanced feature extraction. For this model, hyperparameter tuning was also performed. The results from these tests suggested that the optimal hyperparameters were 20 epochs, and a 0.005 learning rate. Figure 3 displays the segmentation process flowchart.

Our study integrates both classification and segmentation, where classification serves as an initial diagnostic step, and segmentation further refines tumor localization. This dual-stage approach strengthens interpretability, assisting clinical decision-making.

V. Results

The results of this study demonstrate the effectiveness of using deep learning models for both classification and segmentation tasks in the context of breast cancer diagnosis.

In this study, two experiments were conducted using the MobileNetV2 and ResNet50 models with 35 hyperparameter tuning tests performed obtaining varying results. The performance of each model was evaluated based on accuracy, precision, and F1-score across two classes (label 0 and label 1). Table I displays the results of our study.

TABLE I
Summary of Deep CNN Model Evaluation on test data

Algorithm	Best Accuracy	F1 Score	Precision
MobileNetV2	63%	60%	0.61
ResNet50	66%	63%	0.65

A. Classification task results - MobileNetV2

Hyperparameter tuning was conducted to identify the optimal combination of epochs and learning rate for the MobileNetV2 model. The tuning grid included various learning rates (0.000001, 0.00001, 0.0001, 0.001, 0.005, 0.01, and 0.05) and epochs (10, 20, 30, 40, and 50). The objective was to maximize the validation accuracy. The optimal combination identified was 50 epochs with a learning rate of 0.0001, yielding the highest validation accuracy of 0.8333. This model configuration was subsequently saved for further evaluation. The performance of the MobileNetV2 model, configured with the optimal hyperparameters (50 epochs and learning rate of 0.0001), was evaluated on the test dataset. The classification metrics, including precision, recall, F1-score, and support, are presented in Table II.

TABLE II
Classification Report for the MobileNetV2 Model

Precision	Recall	F1-score	Support
61%	60%	60%	27

The multiline plot for MobileNetV2 can be seen in Figure 4. The plot shows that accuracy improves with lower learning rates

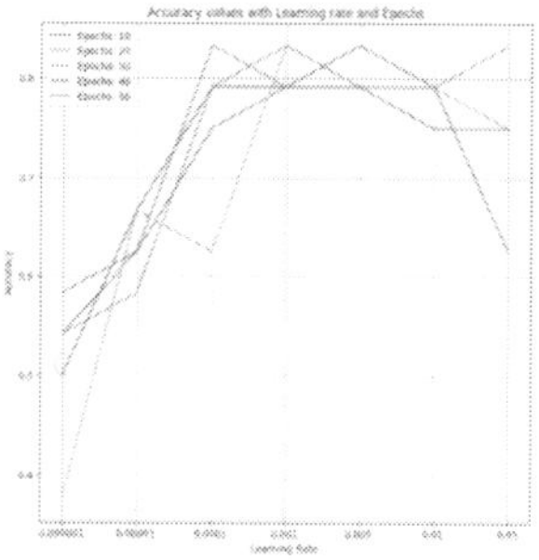

Figure 4. Multiline Plot for MobileNetV2 Model.

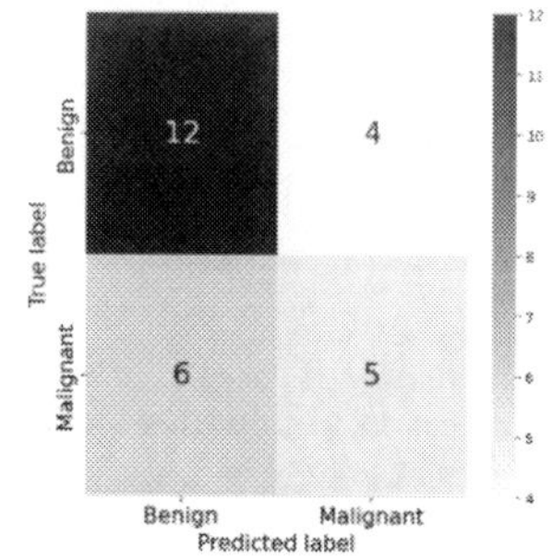

Figure 5. Test Confusion Matrix for the MobileNetV2 Model.

(<0.001) and more epochs, peaking at a learning rate of 0.001 with 30 epochs. Beyond this rate, accuracy declines, especially for larger learning rates (>0.005), indicating training instability. This highlights the need to fine-tune learning rates and epochs, with 0.001 and 30 epochs providing the best balance.

The overall accuracy achieved was 62.96%. For benign cases (Label 0), the model achieved a precision of 66.67%, recall of 75%, and an F1-score of 70.59%. For malignant cases (Label 1), the precision was 55.56%, recall was 45.45%, and F1-score was 50%. The macro average of these metrics indicates balanced performance across classes, while the weighted average reflects performance adjusted by the number of samples in each class. The confusion matrix for MobileNetV2 is shown in Figure 5.

B. Classification task results - ResNet50

Hyperparameter tuning was conducted to identify the optimal combination of epochs and learning rate for the ResNet50 model. The tuning grid included various learning rates (0.000001, 0.00001, 0.0001, 0.001, 0.005, 0.01, and 0.05) and epochs (10, 20, 30, 40, and 50). The objective was to maximize the validation accuracy.

The optimal combination identified was 30 epochs with a learning rate of 0.001, yielding the highest validation accuracy of 0.8333. This model configuration was subsequently saved for further evaluation.

The performance of the ResNet50 model, configured with the optimal hyperparameters (30 epochs and learning rate of

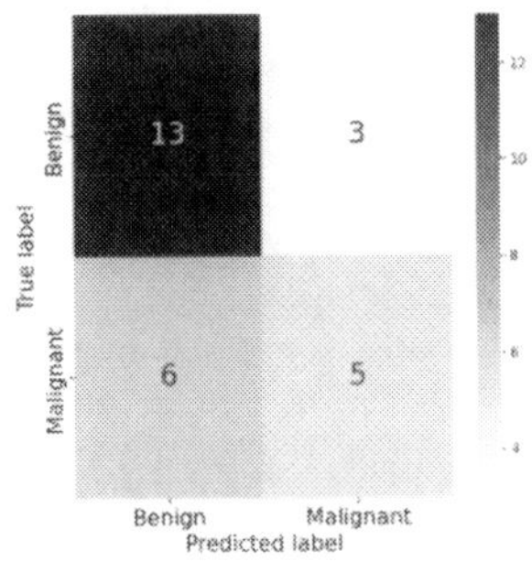

Figure 6. Test Confusion Matrix for the ResNet50 Model.

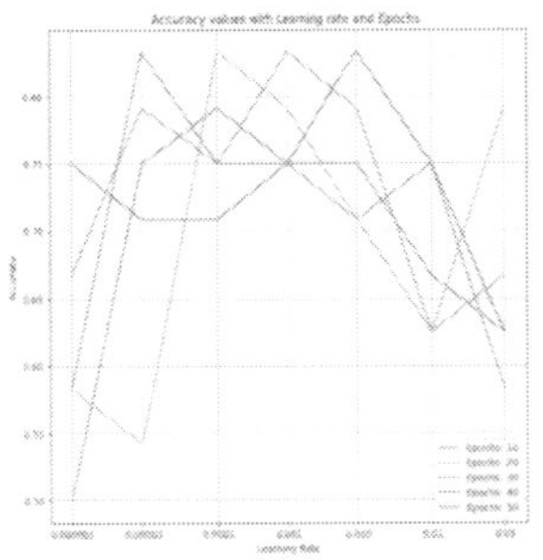

Figure 7. Multiline Plot for the ResNet50 Model.

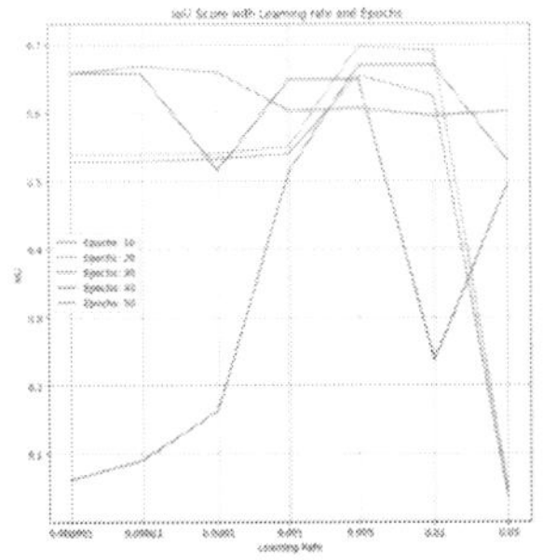

Figure 8. Multiline Plot for the U-Net Model (segmentation).

0.001), was evaluated on the test dataset. The classification metrics included precision, recall, F1-score, and support.

The overall accuracy achieved was 66.67%. For benign cases (Label 0), the model achieved a precision of 68.42%, recall of 81.25%, and an F1-score of 74.29%. For malignant cases (Label 1), the precision was 62.50%, recall was 45.45%, and F1-score was 52.63%. The macro average of these metrics indicates balanced performance across classes, while the weighted average reflects performance adjusted by the number of samples in each class. The confusion matrix for ResNet50 is shown in the Figure 6. As seen, the highest accuracy scores were 0.83, achieved repeatedly throughout testing.

The ResNet50 model outperformed the MobileNetV2 model in overall accuracy, achieving 66.67% compared to MobileNetV2's 62.96%. ResNet50 also demonstrated higher precision and F1-score for both labels, indicating its superior performance in this experiment.

For the ResNet50 model, the classification metrics, including precision, recall, F1-score, and support, are presented in Table III.

Additionally, Figure 7 represents multiline plot for ResNet50. The plot shows the relationship between learning rate, training epochs, and model accuracy. Accuracy generally improved at lower learning rates (<0.001) as epochs increase, peaking near a learning rate of 0.001 for 30 epochs. However, accuracy declined sharply for higher learning rates (>0.005) across all epoch values, indicating instability during training. We can optimize both learning rate and epoch count, with 0.001 and 30 epochs offering a balance between performance and stability.

TABLE III
CLASSIFICATION REPORT FOR THE RESNET50 MODEL

Precision	Recall	F1-score	Support
65%	63%	63%	27

C. *Segmentation task results*

In the second experiment, the segmentation model was tested with 20 epochs and a learning rate of 0.005, resulting in a validation IoU score of 0.697 and test IoU score of 0.629. Table IV shows the summary of the segmentation model's results. Figure 8 shows IoU scores for different learning rates and epochs. This plot demonstrates the IoU score's variation with learning rate and the number of epochs. The IoU score generally increases with learning rates up to 0.001, particularly for 20 and 30 epochs, where the scores peak around 0.7. For higher learning rates (>0.005), the IoU score drops significantly across all epoch values, indicating unstable segmentation performance. The results emphasize that a learning rate of 0.001 and 30 epochs provide the most consistent and optimal segmentation accuracy.

Results from segmentation experiments are shown in Figure 9. The optimal IoU score was achieved at a learning rate of 0.005.

TABLE IV
SUMMARY OF THE SEGMENTATION MODEL RESULTS

Model	Validation IoU	Test IoU
U-Net	70%	63%

VI. DISCUSSION I EVALUATION

The hyperparameter tuning for MobileNetV2 and ResNet50 revealed that the optimal settings for both models resulted in a validation accuracy of 0.8333, but with different configurations (50 epochs and a learning rate of 0.0001 for MobileNetV2; 30 epochs and a learning rate of 0.001 for ResNet50). These settings were selected based on their performance metrics.

In the classification task, the ResNet50 model outperformed the MobileNetV2 model. ResNet50 achieved an overall accuracy of 66.67%, compared to MobileNetV2's 62.96%.

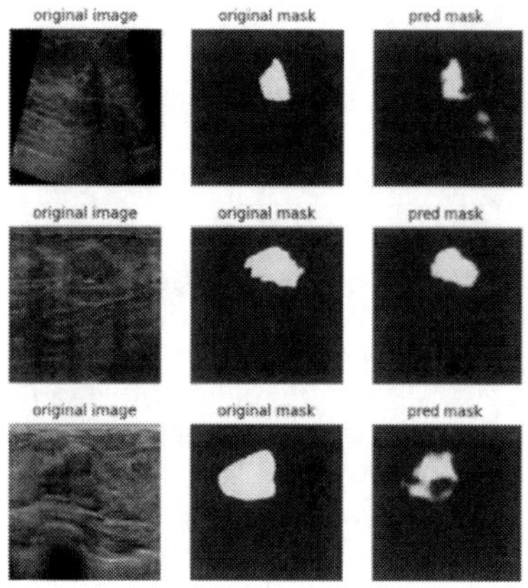

Figure 9. Segmented Mask Images

ResNet50 also showed better precision, recall, and F1-scores for benign cases, reflecting a more balanced and effective classification capability. However, both models exhibited lower recall for malignant cases, which indicates a need for further improvement in detecting malignant lesions.

The results indicate that the MobileNetV2 model is more effective at correctly identifying benign images compared to malignant ones. However, there is a notable trade-off between the precision and recall, especially for malignant cases, which suggests areas for future improvement.

Our results also indicate that the ResNet50 model demonstrates improved overall accuracy compared to the MobileNetV2 model, particularly in identifying benign images. However, the recall for malignant cases still shows room for improvement, indicating that some malignant images are not being correctly identified by the model.

In the application of these models, the differences in false predictions cannot be overlooked. While a prediction of malignant in a case of benign cancer is certainly undesirable, it is especially concerning if a prediction of benign is made in the case of a malignant tumor. Such errors could lead to a lack of testing and treatment for a patient.

For the segmentation task, the U-Net model with EfficientNetB2 as the backbone achieved a test IoU score of 0.629. This result indicates the model's strong ability to accurately segment breast lesions, providing valuable information for tumor localization and quantification. The segmentation results are promising, given the complexity of the task and the variability in lesion appearance.

VII. Conclusion and Future Work

The findings from this study underscore the potential of deep learning models in enhancing the accuracy and efficiency of breast cancer diagnosis. The ResNet50 model's superior performance in classification tasks suggests its suitability for diagnostic applications where accurate classification of lesions is critical. On the other hand, the U-Net model with EfficientNetB2 demonstrated robust segmentation capabilities, which are essential for precise tumor localization and treatment planning.

These results align with existing literature that highlights the efficacy of deep learning models in medical imaging tasks. The use of EfficientNetB2 as a backbone for the U-Net model has shown to be particularly effective in capturing intricate details in ultrasound images, which is crucial for accurate segmentation. The precise segmentation provided by the models means that treatment can be better tailored to each patient's specific condition, enhancing the effectiveness of treatment plans and potentially improving survival rates.

The deployment of these deep learning models in clinical settings can automate and enhance the breast cancer screening process, enabling early detection of cancerous growths with higher accuracy. This early detection is key to improving patient outcomes. By accurately classifying and segmenting breast cancer images, these models can significantly reduce the diagnostic workload of radiologists and pathologists.

Our future research could focus on further refining and optimizing the U-Net deep learning models to enhance their accuracy and efficiency for segmentation tasks, possibly through the integration of more advanced architectures or ensemble techniques. Improvement of IoU score would also be central to development as it provides detailed and precise insights into medical imaging data. Additionally, conducting studies with larger and more diverse datasets would help validate the general applicability of the models, ensuring their applicability across different populations and imaging conditions.

References

[1] Y. Xu *et al.*, "Medical breast ultrasound image segmentation by machine learning", *Ultrasonics*, vol. 91, pp. 1–9, 2019.

[2] S. Radhakrishna *et al.*, "Role of magnetic resonance imaging in breast cancer management", *South Asian journal of cancer*, vol. 7, no. 02, pp. 069–071, 2018.

[3] A. Pawłowska *et al.*, "Curated benchmark dataset for ultrasound based breast lesion analysis", *Scientific Data*, vol. 11, no. 1, p. 148, 2024.

[4] M. Amrane, S. Oukid, I. Gagaoua, and T. Ensari, "Breast cancer classification using machine learning", in *2018 electric electronics, computer science, biomedical engineerings' meeting (EBBT)*, IEEE, 2018, pp. 1–4.

[5] A. K. Singh and B. Gupta, "A novel approach for breast cancer detection and segmentation in a mammogram", *Procedia Computer Science*, vol. 54, pp. 676–682, 2015.

[6] S. S. Almberg *et al.*, "Training, validation, and clinical implementation of a deep-learning segmentation model for radiotherapy of loco-regional breast cancer", *Radiotherapy and Oncology*, vol. 173, pp. 62–68, 2022.

[7] A. Kaur, M. Rashid, A. K. Bashir, and S. A. Parah, "Detection of breast cancer masses in mammogram images with watershed segmentation and machine learning approach", in *Artificial intelligence for innovative healthcare informatics*, Springer, 2022, pp. 35–60.

[8] R. Khaled, J. Vidal, J. C. Vilanova, and R. Martí, "A u-net ensemble for breast lesion segmentation in dce mri", *Computers in Biology and Medicine*, vol. 140, p. 105 093, 2022.

[9] G. Piantadosi, S. Marrone, A. Galli, M. Sansone, and C. Sansone, "Dce-mri breast lesions segmentation with a 3tp u-net deep convolutional neural network", in *2019 IEEE 32nd International Symposium on Computer-Based Medical Systems (CBMS)*, IEEE, 2019, pp. 628–633.

[10] H. Li, D. Chen, W. H. Nailon, M. E. Davies, and D. I. Laurenson, "Dual convolutional neural networks for breast mass segmentation and diagnosis in mammography", *IEEE Transactions on Medical Imaging*, vol. 41, no. 1, pp. 3–13, 2021.

[11] R. Zeleznik *et al.*, "Deep-learning system to improve the quality and efficiency of volumetric heart segmentation for breast cancer", *NPJ digital medicine*, vol. 4, no. 1, p. 43, 2021.

[12] Y. Zhang *et al.*, "Automatic detection and segmentation of breast cancer on mri using mask r-cnn trained on non–fat-sat images and tested on fat-sat images", *Academic Radiology*, vol. 29, S135–S144, 2022.

Determinants of User Trust in an AI-enabled System in the Development Stage

Pi-Yang Weng
Department of Management Information Systems
National Chengchi University
Taipei, Taiwan
email: piyangong@gmail.com

***Abstract*—Explainable Artificial Intelligence (XAI) has provided a noticeable foundation for user trust building in recent years, especially in the high-risk decision scenarios, such as medical and healthcare domains. Building trust in an AI-enabled system is one of the important issues for users, which would start from the development stage. User trust could be enhanced by understanding the so-called black-box model. However, trust could be built by an emotional factor like user satisfaction in addition to scientific factors, such as XAI. In this paper, we present a framework named Three-Pillar User Trust to identify the underlying determinants of user trust in an AI-enabled system. We propose that the introduction of XAI can enhance user trust in the stages of model evaluation and validation by improving their comprehensibility with the AI system outputs and algorithms. Moreover, we propose that user satisfaction, as an emotional factor, would be an important component to influence user trust. To validate our framework, we will recruit some students from one university to participate in our experiment. This research will aim to build a three-pillar user trust framework with model interpretability, user satisfaction, and instance explainability.**

Keywords-XAI; interpretability; explainability; satisfaction; trust.

I. INTRODUCTION

In this research, the AI-enabled system users are the domain experts in healthcare domain, such as nurses or long-term care personnel, in the nursing homes. Recent studies have indicated that AI with explanations allows users to have more confidence in an AI-enabled system and have faith and trust in the algorithm results [1]. In order to obtain a better AI system output performance, domain experts are required to engage in the Machine Learning (ML) pipeline to assist in building an AI-enabled system [2]. It is also important to have domain experts kept in the loop to optimize the ML model [3]. However, ML is a complicated process, especially for deep learning. It is inevitable for domain experts to consider it as a black box even though its inputs and outputs are useful mappings. Therefore, it is essential that an AI-enabled system output is able to be explainable and comprehensible for domain experts to understand, which is instrumental to validate the quality of an AI system output [4]. During the interaction between AI engineers and domain experts in the ML pipeline, domain experts' satisfaction with the AI algorithm interpretation and its output explanation would also influence domain experts' trust in the AI system.

In Section 2, we review related concepts on XAI, Trustworthy AI, and User Satisfaction. In Section 3, we propose a conceptual model named Three-Pillar User Trust. In addition, we propose a research methodology with Hypotheses, AI Artifact, and Experiment Design to validate our framework. In Section 4, we make a preliminary conclusion for this research and propose our future work.

II. LITERATURE REVIEW

The literature review of this research will consist of three parts: Explainable AI, Trustworthy AI, and User Satisfaction.

A. Explainable AI (XAI)

Clinicians might feel uncomfortable with black-box AI, leading to recommendations that AI should be explainable in a way that clinical users can understand [5]. In the machine learning pipeline, users or domain experts are required to participate in model evaluation and system output validation to obtain high-quality training datasets [6]. XAI is a useful tool to unveil the black box and provides an explanation for each AI system output [7], which aims to explain the information behind the black-box model of deep learning that reveals how decisions are made [8]. It is necessary to explain the decision of the AI system to increase the user trust in the system. Therefore, a general model interpretability might not be sufficient for users to build their trust in an AI system. A collection of features to contribute to the output of one specific AI system would be a helpful add-on explanation to enhance user trust [9], which could be defined as instance explainability. Local Interpretable Model-Agnostic Explanations (LIME) [10], one of the XAI tools, will be used in this research.

B. Trustworthy AI

The Defense Advanced Research Project Agency (DARPA) launched a program known as Explainable Artificial Intelligence, whose motivation was to make AI systems explainable and trustworthy [11]. User trust needs to be addressed directly in all the contexts in which AI-enabled systems are being used or discussed [12]. Explainability serves as a fundamental factor that determines the user trust in AI technology [13].

Explainability could be defined as a collection of features of the interpretable domain that have contributed, for a given example, to the production of a decision [14]. To build a trustworthy AI system, a specific instance explainability would be essential for users, especially in the case that the user decision based on the AI system outputs would have a huge impact on its outcomes. (e.g., in the medical and financial domains).

C. User Satisfaction

User satisfaction with the explanation of AI algorithms, which is performed by AI engineers or data scientists could be defined as the degree to which users feel that they sufficiently understand the AI system or the process explained to them [14]. In addition to understanding algorithms in terms of rationality, user satisfaction, as an aspect of emotion, could be an important factor to enhance user trust in the AI system. Recent studies indicated that user interaction with AI-enabled systems would influence user satisfaction with the user-AI system interaction [15]. Therefore, the user would perceive satisfaction with the AI system during the model evaluation and validation while collaborating with AI engineers.

III. RESEARCH METHODOLOGY

AI system users would enhance their comprehensibility with the AI model by incorporating XAI into the model evaluation and validation process. Furthermore, the user comprehensibility would be composed of two components, which are model interpretability and instance explainability, serving as two pillars to support the user trust building. In addition, user satisfaction would be a significant factor in influencing user trust in the AI system. Therefore, we propose a conceptual model as our framework named Three-Pillar XAI for user trust building, as shown in Figure 1.

Based on this framework, we develop our hypotheses and experiment design as follows:

A. Hypotheses

It is essential that the AI system provides users with a reasonable explanation for one instance, especially in a high-risk scenario, such as healthcare domain. Therefore, we develop hypothesis H1 as follows:

H1: Users with understanding about instance explainability would lead to a higher level of trust than users with understanding about model interpretability.

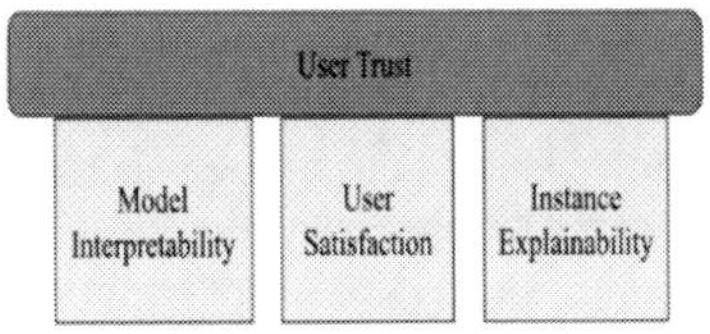

Figure 1. Three-Pillar User Trust.

It is required that domain experts need to be involved in the model evaluation and validation for high-quality training datasets and have a fundamental understanding about the AI algorithm. Then, further build their trust in the AI system. Therefore, we develop hypothesis H2 as follows:

H2: Users with understanding about model interpretability would lead to a higher level of trust than users without any understanding about both instance explainability and model interpretability.

Since user satisfaction with the explanation about the AI system or algorithm would influence his trust in the AI system, we develop hypothesis H3 and H4 as follows:

H3: Higher user satisfaction with the model interpretability would lead to a higher level of trust.

H4: Higher user satisfaction with the instance explainability would lead to a higher level of trust.

We expect that the user trust level with the understanding about instance explainability would be higher than that with the understanding about the general AI model interpretability, especially in the high-risk decision settings. The reason is that users would need to know the reason for one specific system output to ensure that their decision-making is based on logic. Also, we expect that the user trust level with understanding about AI model interpretability is higher than that without any understanding about AI model interpretability and instance explainability. The reason is that users would need to have fundamental understanding about the operational mechanism of the black-box model to build their trust in the AI system. Likewise, we expect that the user satisfaction with model interpretability or instance explainability would be higher than that without any understanding about XAI.

B. AI Artifact

We select the AI-enabled fall detection system as an AI artifact, which is shown in Figure 2. In this research, a mmWave radar is used to detect the moving human body in consideration of privacy, which is a camera-free device. Then, we will use a local explanation tool named LIME to show us the feature importance, such as the speed of movement at different portions of the human body, which indicates the major reason for the fall event and the possible type of fall.

We will show participants the point cloud change in shape, generated from the mmWave radar, while the human body is moving around. Also, we will simulate a fall event to have the LIME generate an output with feature importance.

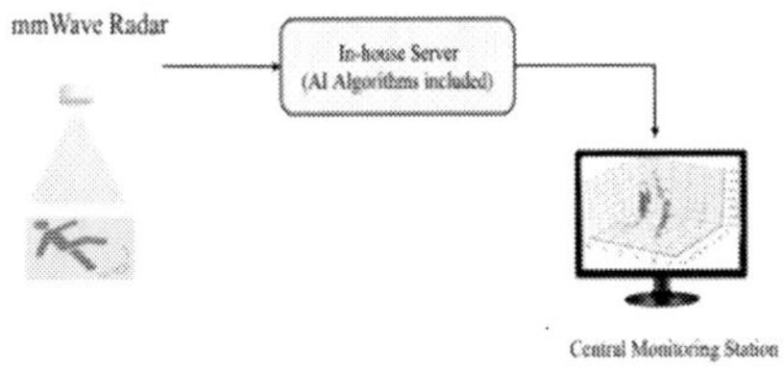

Figure 2. AI-enabled fall detection system with point clouds.

C. Experiment Design

More than 90 students from one university will participate in this experiment and will play the role of long-term care personnel. All students will be randomly divided into three groups, which are group A, group B, and group C. We will design three different courses for different groups, which are described as follows:

Group A: Participate in the model evaluation/validation with instance explainability.

The course outline includes:

- Introduction to fall detection system architecture and functions
- Introduction to model learning process (i.e., ML pipeline)
- Introduction to instance explainability (i.e., system output explanation)

Group B: Participate in the model evaluation/validation with model interpretability.

The course outline includes:

- Introduction to fall detection system architecture and functions
- Introduction to model learning process
- Introduction to model interpretability (i.e., AI algorithm)

Group C: As a control group, without any XAI. Just receive a brief introduction to this AI system, including the system architecture and functions.

IBM SPSS tool will be used for the significance analysis on trust level. In addition, we will check whether the collinearity between these three pillars is not strong, which is required to construct three-dimensional pillars to support this framework.

We design four parts of questions in the questionnaire with 5-point Likert scale, which are partially described as below:

- Model Interpretability
  I understand that the fall detection system uses an AI model, such as the KNN or SVM algorithm.
  I can understand that the change in point cloud shape indicates a certain kind of movement.
- Instance Explainability
  I realize that the AI system will output a reason to show the feature importance for each instance, such as the different moving speed at different portion of the human body.
  I can tell the difference in the human body movement by reading the different feature importance.
- User Satisfaction
  I am satisfied with the model interpretability or instance explainability.
  I think the explanation of the system output is reasonable. (For group A)
  I think the model interpretation is comprehensive. (For group B)
- User Trust
  I realize that this AI system can capture the detecting logic and produce a reasonable output.
  I can rely on the detection result of the fall detection system.
  I can trust this AI system and would like to use it as an auxiliary tool to perform my care work.

Model interpretability could be considered as the first step for domain experts to build their trust in the AI system, providing a general understanding about the AI algorithm. Instance explainability would provide the domain experts with the AI system output reasons. We would anticipate its potential application to expand to a loan application. For example, a bank financial specialist, as a domain expert, may need to know the reasons why an individual loan application will be approved or disapproved, which are generated from the AI system with the capability of instance explainability. Moreover, satisfaction with the model interpretability and instance explainability could be a sense that domain experts perceive the usefulness of the AI system, which is also an important factor for the enhancement of user trust.

IV. Conclusion and Future Work

In this work-in-progress research, we proposed a Three-Pillar User Trust framework based on reviews in the literature, which shows three pillars to support the trust level: Model Interpretability, Instance Explainability, and User Satisfaction. User trust could be built through the user satisfaction with the AI model interpretability or the instance explainability and the user comprehensibility with the AI system output reasons in addition to the user understanding with the AI model interpretability.

User satisfaction is a sense of feeling sufficient and understandable in the AI algorithm and / or system output reason, which is carried out by AI engineers. Therefore, AI engineers would face a challenge in their ability to explain an AI algorithm and the reason behind the output of the system in a way that domain experts can understand.

The introduction of XAI into the ML pipeline would trigger the interaction between domain experts and AI engineers in the collaboration of training dataset generation, model evaluation, and model validation. Moreover, it is a mutual learning process for both domain experts and AI engineers in terms of domain knowledge and ML workflows. Since the result of the model training and the output of the AI system are informed through AI engineers, we might consider it is also an interaction between domain experts and the AI system, which is a human-AI collaboration.

It is possible that this framework could be applied to another high-risk application context, such as the decision on loan application approval. Financial specialists would be highly concerned with recommendations based on the outputs of the AI system because of the huge impact on the consequences of decision making.

Our future work would include more discussions on user satisfaction influenced by the interaction of users and the AI system. Furthermore, we are also interested in constructing an evaluation model for the measurement of user satisfaction.

REFERENCES

[1] D. Shin, "The effects of explainability and causability on perception, trust, and acceptance: Implications for explainable AI". International Journal of Human-Computer Studies, Vol. 146, 102551, pp. 1-40, 2020.

[2] M. Maddi, H. Khorshidi, and U. Ackelin, "A review on human-AI interaction in machine learning and insights for medical applications". International Journal of Environmental Research and Public Health, Vol. 18, pp. 1-27, 2021.

[3] G. Futia and A. Vetro, "On the integration of knowledge graph into deep learning models for a more comprehensible AI: Three challenges for future research". Information, Vol. 11, No. 122, pp. 1-10, 2020.

[4] D. Pedreschi et al., "Meaningful explanations of black box AI decision systems". The Thirty-Third AAAI conference on Artificial Intelligence, pp. 9780-9784, 2019.

[5] M. Ghassemi, L. Oakden-Rayner, and A. Beam, "The false hope of current approaches to explainable artificial intelligence in health care". Lancet Digital Health, Vol. 3, No. 11, e745-e750, 2021.

[6] Google Cloud, "MLOps level 0: Manual process", Google Cloud Architecture Center, 2020. https://cloud.google.com/architecture/mlops-continuous-delivery-and-automation-pipelines [retrieved: December, 2024].

[7] A. Shaban-Nejad, M. Michalowski, and D. Buckeridge, "Explainability and interpretability: Keys to deep medicine. Explainable AI in Healthcare and Medicine", Vol. 914, pp. 1-10, 2021.

[8] A. Chaddad, J. Peng, J. Xu, and A. "Bouridane, Survey of explainable AI techniques in healthcare". Sensors, Vol. 634, No. 23, pp. 1-19, 2023.

[9] M. Rebeiro, S. Signh, and C. Guestrin, ""Why should I trust you?" Explaining the predictions of any classifier". Proceedings of the 22nd ACM SIGKDD International Conference on Knowledge Discovery and Data Mining, pp. 1135-1144, 2016.

[10] D. Kaur, S. Uslu, K. Rittichier, and A. Durresi, "Trustworthy artificial intelligence: A review". ACM Computing Surveys, Vol. 55, No. 2, Article 39, pp. 1-38, 2022.

[11] T. Bach, Khan, A., Hallock, H., Beltrao, G., and Sousa, S., "A systematic literature review of user trust in AI-enabled systems: An HCI perspective". International Journal of Human-Computer Interaction, 40:5, pp. 1251-1266, 2024.

[12] B. Li et al., "Trustworthy AI: From Principles to Practices". ACM Computing Surveys, Vol. 55, No. 9, Article 77, pp. 1-46, 2023.

[13] G. Montavon, W. Samek, and K. Muller, "Methods for interpreting and understanding deep neural networks". Digital Signal Processing, Vol. 73, pp. 1-15, 2017.

[14] R. Hoffman, S. Mueller, G. Klein, and J. Litman, "Measures for explainable AI: Explanation goodness, user satisfaction, mental models, curiosity, trust, and human-AI performance", Frontiers in Computer Science, pp. 1-15, 2023.

[15] C. Rzepka and B. Berger, "User interaction with AI-enabled system,: A systematic review of IS research". Thirty Ninth International Conference on Information Systems, San Francisco, pp. 1-18, 2018.